GW00359389

# The
# Passage Maker's Manual

# The
# Passage Maker's Manual

Bill Finnis

*WATERLINE*

# Dedication

'To all those folk who showed so much spontaneous kindness to us in the course of our voyage and in doing so enhanced those years for us beyond measure'.

Published by Waterline Books
an imprint of Airlife Publishing Ltd
101 Longden Rd, Shrewsbury, England

© Bill Finnis 1993

All rights reserved. No part of this publication
may be reproduced, stored in a retrieval system
or transmitted in any form or by any means,
electronic, mechanical, photocopying, recording
or otherwise, without the prior permission in
writing of Waterline Books.

ISBN 1 85310 440 X

A Sheerstrake production.

A CIP catalogue record of this book
is available from the British Library

# Contents

# Emergency Quick-reference Index

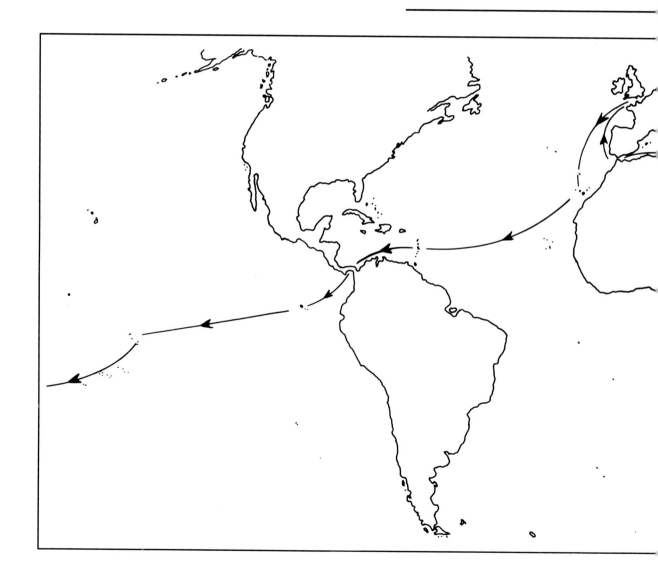

# Route
## 1989

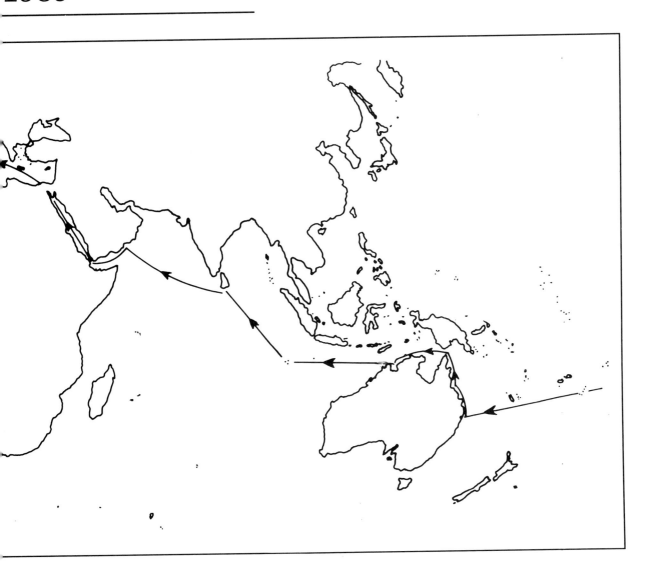

Flying Fish Cove, Christmas Islands, Indian Ocean.

# Introduction

Like so many people, we had dreamed of passage making and over the years had devoured every book on the subject we could find. When we retired we decided that now we had the time, dreaming and reading should be exchanged for the real thing.

Despite many years of cruising, despite all our dreaming and reading, we found to our surprise that we were far from a real understanding of the many special needs, both great and small, of long distance sailing.

Are you aware that, with the exception of North America, flat pack interleaved toilet paper is an unknown item in most of the rest of the world? So our expensive stainless steel flat pack holder had to go into cold storage and be replaced by a somewhat cheaper roll holder.

We could tell a trysail from a storm jib, we knew the difference between heaving-to and lying-to and we knew about streaming warps in bad weather, but no one had told us about flat pack toilet paper.

That may sound like pretty low level inspiration, but for what it is worth it was the spark that produced this book.

In all spheres of seafaring there is more than one way of doing a particular task, but what can be said of what follows is that it has been tried and it works.

# Chapter 1
# The Boat

The purchase of your boat will be your biggest single expense and probably the most important single decision you will make in your preparations so it behoves you to do some research before you spend your money.

You are going to ask a great deal of your boat. It will be your home for several years with all that it demands in comfort and convenience. But you will want more than that. Your boat must be capable of keeping you safe no matter what the sea may throw at you. These two diverse requirements must be kept in mind when you are deciding which boat is to be yours. Comfort may be sacrificed for seaworthiness but never give up seaworthiness for comfort.

The smallest passage making boat we have seen was twenty three feet length overall and the biggest was one hundred and fifty feet, but the average blue-water yacht fell between thirty and forty feet overall. It is a mistake to think that the small boat's gear will also be smaller. Most things, food, cooker, bunks, toilet, charts and much more will be exactly the same size and weight regardless of the size of your vessel.

For six years we were about four inches down on our marks because of the amount of gear we were carrying and this was about normal for most boats we met.

On the other hand the bigger your yacht, the greater the cost of haul-out and anti-fouling paint. This once or twice yearly ceremony cannot be avoided in much of the world you will visit as the tides are far too small to allow you to scrub-off between them.

We found thirty six feet to be a good size, giving adequate storage and living room for the two of us. It is not only length that is important when thinking in terms of accommodation. *Didycoy*'s beam is nine feet which is a little narrow by more modern standards and is made even smaller inside by the requirements of wooden construction. A beamier thirty six footer would accommodate three or four crew, but remember, we are talking of living aboard for several years.

Don't, however, go to extremes in a desire for more room, you could run into stability problems. A beamy, flat-sectioned hull is exceedingly stable up to a point. Beyond that point there is every danger of capsize, after which she will be very stable again....upside down! The old rule that states that the beam should be about one third of the overall length is still valid, the sea does not change.

When we sailed the Great Barrier Reef we met a number of Australian skippers of air-sea rescue launches and pilot boats. These men also act as Queensland's version of the lifeboat service. They were frequently called to the assistance of yachts and fishing boats that stranded themselves on the Barrier Reef. I asked them to comment on the survival chances of boats built of different materials. They were unanimous in their answers.

A steel hull was their first choice. They all said that a well-constructed steel boat will take an immense amount of punishment and still be unlikely to leak.

When we were anchored at the Cocos Keeling atoll in the Indian Ocean we met an Australian couple in a twenty five foot steel boat that had been attacked by a very large crocodile in the Timor Sea. The average crocodile in Australia's Northern Territory is probably ten or twelve feet length overall. The biggest we saw must have been some sixteen feet long and the brute stood at least four feet six inches high at the shoulder. You will appreciate that the sizes are estimates only, mainly for reasons of cowardice.

Even allowing for a measure of exaggeration brought on by a well justified sense of panic, the huge dents and great gashes in the sides of this small boat testified to the fact that it had been attacked by some creature of considerable size and strength. Despite the fact that the boat had been repeatedly attacked, and the evidence was there to be seen, it showed no signs of leaking.

Just so long as a steel hull has been properly treated inside before internal construction reaches a point that makes such treatment impossible, and the exterior is treated in a like manner, rust should not be difficult to keep at bay.

A well constructed timber boat was the second choice of the rescue skippers and when I looked closely at *Didycoy* I could see why. She is built of one inch planking on oak frames which are spaced at six inch intervals. The longitudinal timbers are of considerable size which must lend great strength to the whole shape and the keel and keelson are massive timbers.

One must be realistic and admit that a wooden boat requires a great deal of tender loving care to keep it in good shape. Our previous boat was GRP and the difference in the amount of work needed to maintain the two boats was quite marked.

In tropical waters owners of wooden boats were always conscious of the dangers of a variety of borers simply waiting to eat their boats should they drop their guard. Our hull escaped attack but when we hauled out in Jeddah we were shaken to find that nearly one third of our rudder had been eaten away by teredo worm. The worm had entered the timber through the end grain of the rudder. If I ventured into tropical waters again with the same boat I would certainly remove the rudder and attempt to introduce some very dilute resin into the end grain of the wood and then coat it heavily with resin glue. I think I might be tempted to encapsulate the whole rudder in GRP.

There are resin saturation processes that are reckoned to protect a wooden boat against the depredations of worms and borers. They must, however, be carried out in the course of construction.

Beware of paint on sheathing processes. What can be painted on can just as easily be rubbed off. Some paint on sheathings require that only a soft anti-fouling be used with all its attendant disadvantages.

GRP was the third choice of the Australian rescue skippers. As you would expect, most boats we met were built of glass fibre, none were very old, certainly nothing like *Didycoy*'s twenty five years.

Glass fibre has its own specific problems; loss of gel coat, osmosis and delamination, each one accelerated by the conditions of long distance sailing in tropical temperatures. Continuous immersion and continuous exposure to tropical sunlight for several years is tough on glass fibre. We met a number of yachts suffering from the effects of osmosis and a few whose upper works needed resurfacing.

When we were on passage between Tonga and Australia a vast area of the sea was covered with pumice dust, presumably from an underwater volcanic eruption. The chart for the region was well sprinkled with warnings of underwater volcanic activity. The cartographer left it at that and made no attempt to offer advice in what to do about it. The abrasive effect of wave action by sea water laden with pumice dust working on our boat for twenty four hours a day for several days stripped off our paint above and below the waterline.

A little later we met two glass fibre boats that had crossed the same stretch of sea and they had suffered in the same way as *Didycoy,* but in their case it was gel coat that disappeared.

I am not suggesting for one moment that glass fibre boats were falling to pieces all around the world, far from it. Very few owners indeed neglected their vessels whatever the material from which they were built. Almost all the yachts we met were in excellent condition, so much so that a well kept boat was part of the trade mark of the passage maker.

GRP has many valuable properties, not least are freedom from borers and easy maintenance, but it is not quite the wonder material that we were lead to believe it to be in its early days. Certainly I think it would be unwise to buy an elderly glass fibre boat for passage making.

My informants placed ferrocement boats at the very bottom of their list. It seems that once you have knocked a hole in a concrete vessel part of its total strength is lost and further pounding produces greater and greater damage. For this reason insurance companies in Australia are reluctant to give cover on ferrocement boats.

The rescue skippers said that they put GRP vessels well behind steel and timber boats because they behaved in the same way, although nothing like so badly as concrete boats should they find themselves up on a reef. They declined to comment on aluminium boats as they had little or no experience of them in the sort of situation we were discussing. We too met so few aluminium yachts that my comments would be of no real value.

There is now a whole generation or more of yachtsmen whose experience has been confined to racing boats and racing practice, and not surprisingly, the belief has taken root in some quarters that if it is a good racing boat it is also a good cruising boat. It makes no more sense to use a racing boat for serious cruising than it would to use a blue-water cruising boat for racing. A good sea boat is not simply a failed racing boat, far from it. The requirements for the two activities are poles apart.

A racing boat has two prime requirements, speed and manoeuvrability, and everything else is sacrificed to those two Gods. Neither speed nor manoeuvrability rate very highly on the passage maker's list of essential qualities. Speed, in part, requires lightweight construction and that is no kind of a vessel to be exposed to the unrelenting attack of the sea for several years.

Despite what has often been said about corks floating unharmed in heavy seas, we are not sailing corks. The odd rogue wave that runs across the normal pattern of the seas can hit the side of your yacht with a mind-jarring, tooth-breaking blow that can split open a vessel if it is not sufficiently strong, and where then is your cork similie? We met a couple in the Gran Canaria who had lost their previous boat for just this reason.

These two requirements, speed and manoeuvrability, are demanded within a set of rules which, hopefully, are an attempt in part to instil a measure of seaworthiness into racing design. So often the clever designer circumvents the rules in his search for that extra tenth of a knot and seaworthiness is the first casualty.

C.A.Marchaj, who must surely rank amongst the best brains in yacht design and just about everything related to it, has recently published a book entitled *Seaworthiness, the forgotten factor*. In it he argues that the search for ever more speed under the I.O.R. rules has produced a dangerous breed of boat with a decided tendency to broaching and capsize. He singles out excessive beam, fin keel, separate rudder and dinghy-like sections for particular blame. Marchaj goes on to state that these factors have so influenced cruising yacht design that many can no longer be regarded as seaworthy.

An unseaworthy cruising boat? ... surely that must be a contradiction in terms.

*Didycoy*, in the foreground, sailing from an anchorage in Venezuela.

In forty thousand miles *Didycoy* has never attempted to broach and that has been a great comfort when most of that distance has been sailed in all kinds of weather far from help of any kind. Yet when I have said this to younger yachtsmen their response has usually been tinged with disbelief, so accustomed are they to the pernicious tendency of current designs to broach.

To broach in bad weather is to invite a capsize (see the section on storm management). To have the misfortune to sail a yacht that is prone to broach in a storm that may last three or four days (and nights too, remember) has got to be an exhausting experience. A competent hand will need to be on the helm all the time, constantly alert to try to counter the boat's all too erratic course-keeping. You can soon run out of alert, competent hands in a prolonged spell of bad weather, especially if there are only two of you.

A vessel that does not behave in this way and is equipped with a good wind vane self-steering gear will look after herself and her crew until better times return. All that she will ask for is a reduction in sail and sensible storm tactics. In the meantime you can retire to your bunks and let her get on with it.

The 1979 Fastnet Race storm produced a significant number of boats that capsized and remained inverted for far too long. Not a pastime to be encouraged when you are all alone on some distant ocean. This prompted some research and a large measure of head-in-sand attitude. Designers have continued to produce go-faster boats with scant regard for their seakeeping qualities.

The few investigations that were carried out produced similar results. Basically the findings amounted to two simple facts:-

The fin and skeg underwater hull configuration on the left, and a traditional long keeled cruising yacht on the right.

1. The large beam of many racing boats is designed to enable them to stand up to a great press of canvas. This is fine until they start to heel badly then they are well on their way to a rapid capsize. Couple this with their tendency to broach and you have a specification for a vessel that should never go offshore.

2. Once a yacht with dinghy-shaped sections is inverted it must heel to about 60° from the vertical before it will attempt to return to the normal upright position. An arc of inverted stability of some 120°.

The traditional hull shape will be striving to return to the normal sailing position after about 15° of heel when inverted. An arc of inverted stability of approximately 30°. The hull shape and distribution of ballast that is responsible for this behaviour in a well-designed cruising boat also means that the vessel becomes increasingly able to resist a capsize as it approaches the horizontal position.

I was interested to read that *Jolie Breeze*, the pilot cutter that won some of the first Fastnet races, suffered the same storm that caused so much havoc in the modern racing fleet in 1979. She had a couple of breakages in the course of the storm. Two empty whisky glasses fell off the saloon table and were broken!

Contrary to some opinions a cruiser does not have to be slow. When discussing cruising speeds it must be remembered that in addition to a good stock of fresh and dried food a deep water cruiser would be carrying at least three months supply of canned food, 100 gallons or more of fresh water, probably another hundred gallons of diesel, lubricating oils, tools, materials and spares for every kind of repair, quite possibly a sewing machine, three or four hundred charts, a library of pilot books, radio and light lists, clothing for the crew, a generator and shelves of books and cassettes, and this far from exhausts the list.

Compare this with the stripped out state of most racing boats. Is it really true that dedicated racing types actually cut the borders from their charts to save weight?

Most passage makers are content to make one hundred and twenty miles in twenty four hours. We have done more, one hundred and seventy miles on a number of occasions but it makes no sense to thrash on in this fashion day after day. The gear is stretched to its limits and one is on tenterhooks wondering what is going to carry away first.

In *Around The World in Wanderer Three* Eric Hiscock says, with reference to their passage from Panama to the Marquesas, 'We could not drive the yacht as hard as we would have wished, for with a lumpy beam sea the motion was so quick and jerky when we exceeded six knots and the roaring of the bow wave was so loud that neither of us could sleep properly in our watches below.'

My memory of the days when we were sailing at speeds close to *Didycoy*'s maximum hull speed is one of noise. The vibrating of the sails and the rigging combined with the protesting creaks and groans of the very fabric of her hull sought to outdo the cacophony of the sea. The crash and roar of the tumbling seas added to the cascading sound of water rushing past the hull made conversation, let alone sleep, almost impossible. To endure conditions of this kind day after day, simply to gain a few extra miles, soon becomes a nonsense.

It must be remembered that the average cruising boat carries a far smaller crew than does the normal racing boat of a similar size and this alone demands a good tempered, seakindly boat. Not only must the cruising skipper preserve his gear but the crew must be kept in good shape too and running for long periods at near maximum speed is no way to do that. I have never understood why some people find it so necessary to strive to clip off a day from a passage time that someone else achieved. Conditions are never the same for any two trips over the same route so what does it prove if one vessel takes X days and another takes X - 1 days ?

No, ocean sailing is to be enjoyed and savoured to the utmost and this requires a relaxed attitude. Indeed the relaxed attitude becomes an important part of the pleasure. For a cruising crew to arrive after a passage of two or three thousand miles in a state of exhaustion is almost unknown and certainly says something about that boat and crew.

Tropical cruising will take you to many places that owe something of their beauty to the activities of the reef-building coral polyp. Areas of this kind are often imperfectly charted and part of the pleasure is to be found in conning your boat into some deserted lagoon and finding a spot in which to anchor in perfect solitude. Coral reefs usually rise vertically from deep water to within a few feet of the surface. To feel your way between great clumps of coral in a boat with fin and skeg underwater configuration is asking for trouble as was demonstrated to us some years ago.

We were heading south on the French canals and were in the Donziere Gorge where the current runs at about four knots. With the five or six knots we were making through the water our speed was nine or ten knots over the ground. Quite suddenly *Didycoy* lifted some way out of the water and adopted a considerable angle of heel but continued to move forward at a high rate of knots. We had climbed onto an unmarked slab of rock that protruded into the channel. We covered about thirty yards in this manner and then fell off at the down stream end of the slab and resumed our normal upright position. The only damage was to the barnacles on the underside of our full length keel, and perhaps to our hearts.

When we reached Port St. Louis at the southern end of the Rhone we berthed close to a boat that was being delivered to the Mediterranean from Britain. This yacht had met up with the same slab of rock in the Donziere Gorge as we had tobogganed over.Unfortunately for them the boat had a deep fin keel. It had struck the leading edge of the slab of rock and the bottom of the boat had been partially ripped out. Because they were in a canal they were able to reach the bank before they sank. Because they were in Europe they were able to organise assistance and repairs. Because they were insured they could pay for the repairs. If the same thing happened to you in some distant coral-studded lagoon the result could be very different indeed.

This view of *Didycoy* from above clearly shows shows her double ender design.

In the light of our experience my choice of boat would be about thirty six feet over all with an underwater shape and a ballast ratio similar to a twelve ton Hillyard but with a twelve foot beam. The material would be substantial glass fibre, chosen for its ease of maintenance, freedom from the possibility of damage by borers and its lack of rust. In addition my ideal boat would have a deep centre cockpit and a doghouse to provide some shelter from the tropical sun.

There is a long standing belief that a double ender is unlikely to be pooped by a following sea. It is a statement that I never really accepted. It seems to me that most boats are double enders at the waterline. *Didycoy* is a double ender and she has been pooped twice. Conditions were rather bad on both occasions and I was less than happy to be proved right!

The first time we were pooped we were about fifty miles west south west of Gran Canaria and so much water came aboard that the after end of *Didycoy* was totally submerged. Fortunately the after end of the doghouse was closed off with the tonneau and a very large measure of water shot over the doghouse and cascaded onto the forward coachroof without coming into the cockpit. For a while *Didycoy* felt like a half tide rock but she shed the water, shook herself free and sailed on. An after cockpit would have filled, completely. We chose a centre cockpit with the possibility of pooping in mind and we were very glad that we had done so.

When we had a close encounter with a reef in the San Blas islands we were repeatedly laid over horizontally and completely washed over by pounding surf. Our doghouse and our deep cockpit saved our lives. There is no question in our minds that we would have been washed out or thrown out of the average shallow dishpan of a self-draining cockpit.

If in a future boat I can have a deep self-draining cockpit well and good, but if it must be one or the other I will leave the self-draining cockpit to others and put my faith in adequate bilge pumps.

## Non-slip Decks

Coral sand and ordinary paint make an excellent non-slip deck, anything but very fine sand will be painful to kneel on.

Paint an area about three feet square and immediately cover it liberally with fine sand. Leave the edge of the area free of sand or you will have difficulty in joining on the next painted area. Continue in this fashion until the whole area is covered.

Twenty four hours later brush off all the loose sand. Use a soft brush or you will pull off much of the sand that is adhering to the paint. When all the loose sand is off give the area another coat of paint and the job is done.

The salt content seems to make no difference but it is important to use dry sand. I punched a number of holes on the bottom of a can and sieved the sand through that simply to get a fairly uniform particle size which makes for a better looking job.

Don't be tempted to add sand to the can of paint. Mixing sand and paint is a tedious chore and painting with it is even worse.

The process is heavy on paint, needing enough for three coats to achieve two. The sand comes free.

Landfall off Tenerife - on the way out.

# Chapter 2
# Rigging

Trade wind sailing is one of life's unique experiences ... it is even better than you expect it to be. Our first taste of the trades was our passage from The Canaries to Barbados. It took us three days to find the trade winds and then we were off and away. Twenty four days of superb sailing and sunshine with little need to touch a rope. It was unbelievable. For all this time the wind was on our starboard quarter at a steady 20 knots day and night and we goose-winged our way across the Atlantic as if we were on a magic carpet.

**Preventers**
Just occasionally a wave that is out of step with the rest will throw the boat far enough off course to cause the sails to be taken aback. If they are not properly secured the vessel would gybe all standing and this could be dangerous for gear and crew alike.

To guard against a gybe in these circumstances both the boom and the whisker pole must have a vang and a fore guy preventer. Note the word 'preventer'. It is not there to cushion the effects of an unintentional gybe but to prevent it happening. From this it follows that the preventer must not be fashioned from nylon but from something like prestretched Terylene that will not allow the spar to move should the sail be taken aback.

Deck gear of this kind needs to be on hand ready for use, not stowed in a locker in the cockpit. It should be possible to operate this gear from a safe position on deck and the safest position on deck is between the mast and the shrouds, preferably up against the shrouds.

Our main boom preventer is in two parts. A length of line is shackled to the outboard end of the boom and brought forward to the goose neck so that it lies close to the length of the boom. The forward end of this line ends in an eye splice and when the preventer is not in use it is attached to a hook at the goose neck.

The other half of the preventer runs from a cleat on the toe-rail close to the shrouds, forward over the deck, through a block at the stemhead and then outside the guard-rails to the shrouds where it ends in a snap shackle that is clipped onto a shroud at guard-rail level.

In use the eye splice is unhooked from the goose neck and the snap shackle is removed from the shroud, the two ends are married up, snap shackle to eye splice, outside the shrouds. The line is then hauled taut and secured to the cleat that is fixed to the toe-rail close to the shrouds. Note that the operation is carried out between the shrouds and the mast and boxed in on one side by the boom.

The second half of the preventer must be duplicated on the other side of the boat for use on the other tack.

A preventer does not need a tackle to help haul it taut. When it is secured a pull on the main sheet will make everything as tight as you could want. If the preventer is used alone and the mainsail is taken aback, the boom could rise to quite a steep angle and possibly clash with the backstay or hook the mainsail around the cross-trees. To ensure that this does not happen a vang must be rigged. It may be that your boat has a kicking strap that will serve, particularly if you have a short boom, if so, well and good. If it does not then a vang must be rigged.

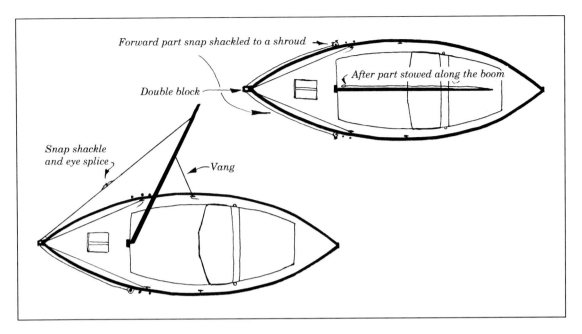

*Forward part snap shackled to a shroud*

*After part stowed along the boom*

*Double block*

*Snap shackle and eye splice*

*Vang*

Fig 1. A diagram of the Preventer system used on *Didycoy*

**Vangs**

If you have a long boom it may be necessary to anchor the lower end of the vang in the region of the toe-rail to get the support far enough along the boom. A vang must rise towards the outboard end of the boom. It should never rise vertically, or worse, towards the goose neck, this would be a recipe for a broken boom.

*Didycoy*'s vang lives on the underside of the boom with one end attached about half way along the boom and the lower end of the vang snap shackled to a becket at the goose neck. The vang is a three part tackle and when it is needed the snap shackle is taken from the goose neck and hooked to a rope eye on the toe-rail close to the shrouds. The fall leaves the lower block giving an upward pull which is easier to use than it would have been if it had come from the block on the boom. Whilst it is true that the tackle is used to disadvantage by hauling from the lower block, the load is not excessive and the convenience of hauling upwards more than makes up for the loss of some mechanical advantage.

Because the vang must not be used as a preventer, particularly on a long boom, it should be rigged after the preventer is in place and it must be stowed before the preventer is removed. Note that once again the work is carried out in the area between the mast and shrouds.

**Whisker Pole**

The whisker pole that is used to extend the jib when sailing with the wind abaft the beam can be a lethal instrument if it is not installed properly. It may be in order for an overmanned racing boat to stow its whisker pole at deck level and then to have a small army of tame muscle men descend on it when it is needed. A husband and wife cruising team must operate in a different way if they are to escape injury or even death by drowning.

Here the vang can be seen pulling the boom down, with the tackle clearly shown. The preventer is the line running from the boom towards the bottom left of the picture.

A deck stowed pole has jaws at both ends, the inboard end is intended to clip onto an eye that is housed in a short length of track on the forward face of the mast. For a short handed crew in a seaway this alone can prove to be a daunting task. When that is accomplished a topping lift and the clew of the jib must be attached. For a man alone on the foredeck in a boat that is literally rolling its way down a trade wind route the whole process is fraught with difficulty and danger. When the pole is no longer needed the whole painful process must be repeated in reverse order.

For a cruising boat, short handed or not, there is only one safe way to handle the problem and that is to stow the pole up the forward face of the mast on a purpose-made track.

In this system the inboard end of the pole terminates in a strong slide fitting that incorporates a joint allowing movement in the horizontal and vertical planes. This slide is raised and lowered by a continuous line that acts both as a halyard and a downhaul. The outboard end of the pole has a fixed topping lift, the effect of which is to cause the outboard end to rise to the horizontal as the inboard end is lowered. It must be clearly understood that the standing topping lift is an essential factor in the safe operation of this rig.

When the pole is needed, the downhaul is used to lower the inboard end far enough to cause the outboard end to rise under the influence of the topping lift to guard-rail level, where it rests on the rail against the shroud. With the pole lying quietly in this position it is a simple matter to clip the jib sheet into the jaws. When the sheet is in place the inboard end of the pole must be lowered to bring the pole to the horizontal position and then the jib is ready to be raised, or if you have a roller jib, extended.

The mast attachment of the permanently rigged whisker pole. The inner end can slide up and down the track by using the control line and when stowed upright it rests at the bottom of the track.

## Topping lift

The topping lift has been referred to as a fixed topping lift which is not quite accurate. The fall of the topping lift passes through a jamming cleat that is screwed to the mast. The intention is to give some very limited variation in the length of the topping lift. This will allow the outboard end of the pole to be raised or lowered a little to accommodate the different positions of the clew for a variety of jibs. To all intents and purposes the topping lift is fixed and it is a most important part of the system. Two lines are shackled to the outboard end of the pole and when the pole is rigged these lines are secured, one aft and one forward to prevent an accidental gybe.

Whilst any adjustment is being made to the position of the pole, the jib sheet must be slack otherwise the pole will not be free to move and adjustment will be impossible.

With the sheet through the jaw of the pole it becomes possible to get the jib sheet home before the sail is raised which is preferable to fighting a thrashing jib to a standstill.

During the rigging or stowing of the pole with this system the whisker pole is under total control throughout the whole operation which is more than can be said for some other methods to which I have been subjected.

In the course of my sailing career, whisker poles have developed from cheap, simple items of equipment to heavy, expensive high tech telescopic pieces of gear with decided antisocial tendencies. I speak with some feeling, having recently had two sizeable slices of flesh torn from my fingers by one of these over-rated gadgets.

*Didycoy*'s whisker pole is an ancient wooden spar, two and a half inches in diameter. We have used it for forty thousand miles, most of which was down-wind work. How far

13

The rigged whisker pole in action. The preventer can be seen running down to the bow and the sheet runs aft from the outer pole end. The topping lift is also visible running upwards from the outer pole end.

previous owners sailed with it we have no way of knowing, but it must be fair to say that it has earned its keep and it will be so easy to replace should the need ever arise.

A whisker pole is nothing more than a strut and wood is at its strongest when it is in compression. It need not be circular in section and what is more it will float should it be dropped over the side, which is more than you can say for an aluminium pole. As for the telescopic facility, the need for it disappears as soon as the jaw of the pole is placed on the sheet rather than in the clew of the sail. Unless there is something badly awry with the jaws of your whisker pole the jib sheet will suffer no chafe.

There is a point of sailing where the shrouds prevent the use of a whisker pole. Usually this point arrives on a broad reach when the jib needs to be poled out on the same side as the mainsail. To overcome this problem the jib sheet passes through a block at the outboard end of the boom, down to the deck where it runs through a turning block and then leads to the winch in the usual way. When it is not in use the sheet we use for this purpose lives on the underside of the boom, reeved through the block at the end of the boom.

In use, the end of the sheet that has the eye splice in it, is taken outside everything and fixed to the clew of the jib. The other end goes via the block at the outboard end of the boom to a turning block at deck level to give the sheet a rising approach to the normal jib sheet winch. With the jib sheeted to the end of the boom in this fashion, it also allows us to goose-wing two jibs should we wish, one on the pole and one on the boom end.

The Dutchman's method of attaching sheet to headsail. The line with the eye splice and manrope knot is inset top left.

### Jib Sheet

Some years ago I noticed that Dutch yachtsmen did not waste their money on fancy stainless steel gadgets to unite their jib sheets with their jibs. Instead they used a short length of rope with an eye splice in one end and a manrope knot in the other. I tried it, liked it and have not used anything else ever since. So far it has never let me down.

In use the eye splice is passed through the eye of the sheets and the clew of the jib. The eye splice should be just big enough to allow the manrope knot to be pushed through it. A figure of eight knot would serve in place of the manrope knot but perhaps not look quite so handsome. It is not unusual to use a bowline for this purpose but a few days of hard sailing so often tightens the knot to a point where it can be very difficult to release.

### Sail Bags

We have seen some boats with what I can only call pouch-shaped, standing sail bags. There are usually two of them, one fixed inboard to either side of the guard-rails right forward so that a lowered sail can be stowed directly into one of them and left in situ. They are used in conjunction with twin forestays and the piston hanks are usually left clipped to the stay. The top of the bag is so shaped that the piston hanks are covered by the top flap of the bag and so protected against the damage created by exposure to strong sunlight.

I have no personal experience of these sail bags because we stow our dinghy on the foredeck and it does not leave room for other bulky gear. If we had room I would certainly make a pair as they have all the hallmarks of a good idea. Certainly the reports I elicited from those who used them were very favourable.

## Twin Forestays

If you have twin forestays, and I am glad we do, make sure that they are well separated. If they are too close together you will find that the unoccupied stay will often catch some of the piston hanks and open them, so freeing the jib from the stay. It is not unknown for the freed piston hank to re-attach itself to the forestay. What is unknown is for the freed piston hank to re-attach itself to the correct forestay. When several hanks have transferred themselves in this way to the wrong forestay in the course of the night it can be a hard and long slog to lower that jib the next morning.

The only certain cure we found was to hoist the jib on the leeward stay. By this means the wind pulled the stays apart rather than towards each other. Maybe this sounds like a lot of bother but remember that we are talking in terms of sailing on that one tack for several days, maybe weeks even.

## Main Booms

The current practice of working on an unsecured boom is the height of folly; I can think of no simpler way to arrange to be thrown over the side. Boom gallows have been outlawed in the search for that extra tenth of a knot by those who race and the rest of the sailing world, without a thought, has followed their lead, as it has in so many things. A gallows does not have to be a cumbersome thing. A piece of stainless steel tubing that is also part of the spray hood is one possibility or if your boat has a doghouse, a small, shaped block of wood that is fixed to the top of it could be the answer.

Whatever form it takes a strong boom gallows is an essential piece of equipment if you intend to go to sea. It takes a large measure of the difficulty and danger out of reefing or stowing the mainsail in any sort of seaway. Once the boom is secured to the gallows it is transformed from a lethal weapon to a solid handrail.

Bring your topping lift down the backstay instead of the mast and it will be possible to house the boom in the gallows from the safety of the cockpit. An involuntary gybe will lift the boom end quite high, so producing a great bight of slack line in the topping lift which can trap the boom with rather uncomfortable results. To avoid this you must seize a stainless steel ring to the backstay three or four feet above the level at which the boom rests in the gallows. With the topping lift halyard led through this ring it is unlikely to foul the boom end should you gybe all standing. At deck level the halyard must be taken through a turning block and then to a cleat in the cockpit.

I had read somewhere that the topping lift should be a nylon rope. We followed this suggestion and cursed the author all the way across the Atlantic. Every time we tried to raise the boom we filled the cockpit with the fall of the nylon topping lift and the boom obstinately refused to rise.

## Lazy Jacks

Lazy jacks. God's gift to short handed sailors. Like gallows, lazy jacks are so much a thing of the past that many yachtsmen have never heard of them. In use, when the main halyard is released the sail drops to the boom, restrained by the lazy jacks, and remains there without engulfing the helmsman. They help restrain a partially lowered sail when a reef is being tucked in, a time when any help is most welcome. What is more the mainsail can

The boom laying safely in the gallows. At sea the security of a safely
contained main boom can be a life saver.

be dropped and forgotten whilst you are busy anchoring. The mainsail can be so secure there that we have often left it over night with no additional lashing.

Quarter inch diameter line is adequate for lazy jacks in most boats. A line is fixed to either side of the mast well above the crosstrees. The lower end of each of these lines terminates in a small block well above the centre of the sail. A continuous line is then reeved through the blocks passing under the boom as shown below.

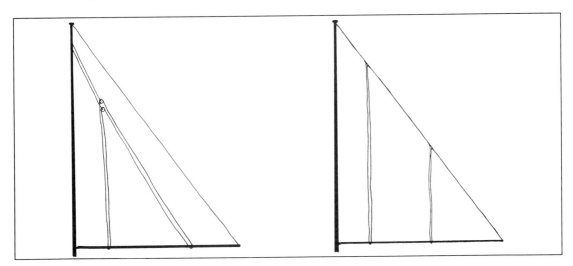

Fig 2.The system of rigging lazy jacks.

It will probably be necessary to make some provision to retain the lines in the places where you want them as they pass under the boom. Lacing eyes screwed onto the underside of the boom will serve unless you intend using roller reefing. I must say that if roller reefing denied me the use of lazy jacks, I would prefer a form of reefing that allowed me to use them.

There is now on the market a lazy jack called a 'Dutchman' that is possibly a little better even than the lazy jacks I have just described. It requires one or more vertical lines of small reinforcing patches to be sewn into the mainsail, each with an eyelet set in it. Each vertical line of eyelets has a line threaded through it. The top end of the line is spliced to the topping lift and the lower end is fixed to the boom. In use as the mainsail is lowered so the lines cause it to flake down on the boom, alternately to port and starboard. The Dutchman's one drawback is its price, it is very expensive.

## Reefing Points

Initially our mainsail was furnished with a lanyard instead of reefing points and it was always difficult to manage in strong winds — the only time it was needed of course. I could never get each part of the sail secured in a uniform manner and the end of the long lanyard was for ever being snatched from my hand by the wind as I worked.

At the earliest opportunity we changed to reefing points. The change was not difficult. I passed suitable lengths of line through each eyelet that had been provided for the lanyard, holding them in place with an overhand knot on either side of the sail as close to the eyelet as it was possible to get.

Should you decide to do this, be generous with the length of line that you use for your reefing points. It makes reefing much easier if the points are long rather than short. I used three strand line for one set of reefing points and bright red braided line for the other set. This proved to be a great help in distinguishing between the two rows of points when I was struggling to get a reef tucked in. Even in the dark it was possible to feel the difference in the two materials.

If you are having a mainsail made make sure that the reefing points rise as they progress from luff to leech. By this means the boom end will be lifted higher as the weather deteriorates which will help keep the end of the boom out of the water.

## Reefing Pendants

Each row of reefing points ends in a reefing eye at the leech of the sail. A reefing pendant is spliced into each of these eyes from whence it descends to the boom, where it is turned by a Bee block and leads along the boom to the forward end where it is cleated. The end of the pendant should go through a hole in the cleat where it is retained by a stopper knot. The purpose of the pendant is to pull the leech of the sail down to the boom and hold it there. For this purpose the line needs to be strong but if it is too heavy it will be forever pulling the leech of the sail down and spoiling its shape.

It is possible to use a comparatively light line for the reefing pendant which helps avoid distortion at the leech, but is then far too thin for comfort when it is in use at the forward end. I overcame this by using two diameters of line, a thin one for the vertical part and a heavier line for the horizontal part, joining them at the boom with a splice that had a substantial seizing to reinforce it.

**Reefing**
Our first step in reefing is to head close to the wind, get the boom into the gallows and the mainsheet made up so that I can rely on the boom to help keep me inboard.

The main is then lowered rather further than is needed to get the line of reefing points onto the boom. If you don't create enough slack at this stage you will make difficulties for yourself in the subsequent steps. We have two large hooks welded to the goose neck which are intended to take the reefing eyes on the luff of the main and the next step is to use the appropriate one for this purpose.

Once this is done the correct reefing pendant is used to haul the leech of the sail to the boom and keep it there. A small winch is bolted to the underside of the boom to help with this step.

So far the work has been done in the region of the mast which will give you a degree of security. The next step requires a move onto the coach roof where you will tie the pairs of reefing points under the foot of the sail. I keep a four foot long piece of line attached to my safety harness which serves to fasten me to the boom and makes it possible to use both hands for reefing.

When I reach the leech of the sail, I tie the reefing eye to the boom with several turns of light line to back up the reefing pendant. Should the pendant be used alone it could part under the load imposed on it and if that happened the main would probably be ripped across the line of the reefing points.

It remains to rehoist the mainsail until the luff is taut and remove the reefing pendant from the winch barrel so that the winch is free for the second reef if it is needed.

When shaking the reef out, the reefing points must be the first lines to be released otherwise the sail is again exposed to the possibility of being ripped across.

The tails of all halyards must be secured at deck level. It is an immense help to be able to uncleat a halyard and forget it, knowing that the fall will not be flying at the masthead when you have subdued the sail you are lowering.

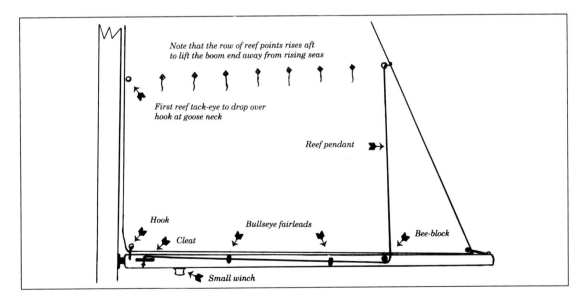

Fig 3. Mainsail reefing system used on *Didycoy*.

## Chafe

Chafe is an ever-present enemy that can damage or destroy gear so quickly that it is unbelievable.

When under way and a sail is lowered, remove the halyard from the peak of the sail and secure it. If it is left attached to the sail the weight of the snap shackle will cause it to swing back and fore and produce a damaged area of sail in a very short time.

Jib sheets will rub against a guard-rail on some points of sailing and twenty four hours of this treatment will be more than enough to start the ruination of a good pair of sheets. Why use a pair of sheets, why not a single sheet? If the jib is to stay on that tack for several days there can be no reason to expose the unwanted sheet to days of sunshine and chafe. If it's not rigged it can't be damaged.

Plastic tubing will go a long way towards eliminating chafe. We first tried to protect our sheets by sliding a five foot length of plastic tubing onto each sheet. It was possible to move the tubing along the sheet to cover the point of contact. It worked well enough but with one snag. In the flurry of sheeting-in the tube would sometimes foul the turning block so we tried placing it on the guard-rail. This was the better option.

If you are to place plastic tubing in strategic places on your guard-rails it must be large enough to rotate as the sheet bears on it. To get it on the rail, the tubing will need to be split with a Stanley knife and then when in place it must be held there with a short length of French whipping at either end. Don't be tempted to use insulating tape, it will not last more than a week or two before it starts to come unstuck.

The same treatment can be accorded to the lower twelve feet or so of your shrouds. This will give your sails and sheets a large measure of protection from chafe. I know that baggy wrinkle gives a raffish air to a cruising vessel but it is a tedious item to make. Look around stores that sell plumbing and electrical gear. The chances are you will find cheaper and more suitable plastic tubing in one of those places than ever you will in a chandlers.

The outboard ends of your cross-trees must be well padded to prevent damage to your foresails.

## Harnesses and jackstays

After some years experience of *Didycoy*'s chest-deep cockpit, I find it somewhat disturbing to sit in a modern self-draining cockpit, mainly because they are so shallow. If I ever own a boat of that kind, the first thing I shall do is install strong points in the cockpit so that I can clip myself on should the weather start to deteriorate; call me a coward if you wish. I suspect that the most dangerous moment in this kind of cockpit is as one is emerging from the cabin in lumpy conditions. It should be possible to clip-on before leaving the security of the cabin.

A wire jackstay at deck level, running from the cockpit to right forward on either side of the boat on which to clip your safety harness is a must. If yours is a centre cockpit boat then a further pair of jackstays should run aft from the cockpit.

I have never been completely happy with the standard pattern of the safety harness. Many are clumsy and difficult to put on in a hurry and some are still sold with the simple carbine hook that was proved to be utterly unreliable many years ago. So when I learned that only one harness offered for sale at that time had the approval of the British Standards Institute I decided to make my own.

After one or two false starts I arrived at the pattern that is illustrated. I know it works well, as it is the harness I was wearing when I went overboard in the Caribbean. I went

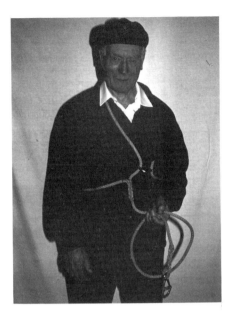

My own answer to the safety harness. It works!

over the side in fine weather in the Caribbean and it is probable that I would not have been picked up if I had not been wearing a clipped on safety harness. I can still recall thinking, as I flew through the air, 'Well, if it doesn't work I have no one to blame but myself'. It is simple to make, easy to put on, you can be sure of its effectiveness and what is more it is cheaper than most of the commercial models.

Work a large eye splice into one end of about 3½ metres of ½in diameter soft rope. The splice must be big enough to allow you to place an arm and a shoulder into it. Captive on the rope of this eye splice is a strong clip hook, which can be of the forbidden pattern because in this position it cannot be opened accidentally. Before making the eye splice, experiment with a bowline to find the right size bight for yourself.

A small eye splice must be worked into the other end of the line and captive within it is the best safety harness clip you can find. It won't be cheap but make sure that you get a clip that is intended for the task and will lock in the closed position.

Our self-steering rudder made an excellent climb-aboard point. If your boat lacks this facility you must provide some means of getting back inboard because without it there is no way you are going to pull yourself back over the toe-rail.

**Mast Ladders**

I have tried to make a point of climbing the mast after every two or three thousand miles to examine every fitting for damage or deterioration. On these inspections I have found a number of actual and potential problems that ranged from a nuisance to positively dangerous.

We have steps from the cross-trees to the top of the mast. Below the cross-trees we were forced into having ratlines because our storm trysail is secured around the mast with heavy lanyards which would never have got past the steps. If I had a free choice I would install steps all the way to the masthead.

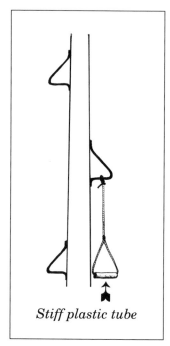

*Stiff plastic tube*  Fig. 4

Don't be tempted to economise by placing the steps too far apart. Climbing a mast is no fun and it is not improved by having to stretch at every step. At about four feet from the top of the mast you should fix a step on both sides, so that you can stand there and work on the various items of gear that live there. If you are working at any point but the masthead you will be standing on one foot on a narrow strip of metal, and it rapidly becomes anything from uncomfortable to painful. The temporary stirrup illustrated in Fig 4, clipped to the next step up on the other side allows you to support your weight on two feet instead of one.

Whenever I climb our mast I wear a wide safety belt to which I can clip a halyard. It has a further belt sewn to it that ends in a safety harness hook and a stainless steel ring. The additional belt is long enough to pass around the mast and allow me to lean back and stay in position without holding on. In this way I have both hands free to work.

Mast steps, especially those above the cross-trees, tend to ensnare halyards. To avoid this happening, secure a length of light line from each step above the cross-trees to the nearest cap shroud. Use a separate line for each step. If you use a continuous line for all the steps on one side of the mast, one chafed spot will mean renewing all the line instead of one short piece.

Originally I tried rope rungs on our ratlines, but they pulled the shrouds together in an alarming manner as soon as I started to climb them. I then resorted to one wooden rung followed by two rope rungs. It was better but not really good enough so it had to be timber slats all the way up.

A yacht's shrouds are so thin that they don't provide enough body on which to lash a rung effectively. This can be overcome by fixing bulldog grips to the shrouds as points on which to rest the rungs prior to seizing them to the shrouds. Until I did this I was forever relashing ratlines that had slipped out of place.

## Standing Rigging

The standing rigging of most racing yachts, and many inshore cruisers too, is quite inadequate for passage-making. Racing boats are seeking to reduce windage and weight, so they resort to the lightest rigging wire they think they can get away with. The result is seen whenever the racing fleets are subjected to a anything above a force five.

That masts should fall down is accepted as a hazard of racing. What is an inconvenience and an expense to a racing boat could be a major disaster for a blue-water cruiser and so higher standards must be observed.

All passage-making boats are overladen because of the very nature of their undertaking, so the few extra pounds contributed by using adequate rigging is of no consequence. As for windage, I have calculated that to increase the diameter of all of *Didycoy*'s standing rigging by one eighth of an inch would increase the area exposed to the wind by about three square feet, roughly equivalent to half a crew member, hardly something to get uptight about.

For a vessel that is to withstand the rigours of offshore, sailing the angle between a cap shroud and the mast at the masthead should be between 12° and 15°. Anything less will not give proper support to the mast. For reasons best known to their designers, many modern boats have an angle of only 8° in this region. If your vessel has its cap shrouds set up in this way, wider cross-trees might be the answer.

When the sails are down and stowed, the halyards need to be frapped so that they do not keep you and your neighbours awake in anything of a breeze by strumming against the mast all night. Easier than tying them back to the shrouds is to catch them behind thumb cleats that have been fixed to the cross-trees.

Rod rigging, hydraulic backstays, forestay adjusters, mast rams and benders, boom benders and stretchers and the rest are all unwanted sources of trouble in a seagoing vessel.

Note the lower ratlines with solid wooden steps. Also the main boom vang, preventer and the jackstay running along the deck.

Chapter 3
# Sails

There is an air of mystique associated with the work of the sailmaker which makes many people shy away from attempting their own repairs. I think that it is the designer of sails who perhaps deserves the aura of mystique rather than the whole profession, which is not to say that people at lower levels don't have skills and knowledge that has been hard won. I would be the last person to deny a craftsman due recognition for his skills, learning and experience, but your friendly neighbourhood sailmaker is not going to be on call I am afraid and, as in so many other areas, you are likely to have to depend on your own efforts.

Even when you can find a sailmaker ashore in some remote spot, he may well be a sailmaker in name alone and you will not know that, until he has worked on your sail. I have repaired numerous sails for other yachties, many of which had been tackled by operators ashore who had no understanding of yacht's sails. This always made the salvage operation a bigger job than the original repair would have been.

If the sailmaker in some out of the way place does not have a machine that will produce a zig zag stitch then you must look elsewhere, or at least insist that he uses a really large stitch.

Continual exposure to tropical sunlight hardens man-made sail-cloths and when this has happened every new stitch will damage the fabric. Zig zag stitching will stagger the line of stitches so leaving more undamaged fabric between stitches than would a single straight line of similarly spaced stitching. If the line of sewing is straight and the stitches are small the row of needle punctures creates a terrible line of weakness that will tear at the first opportunity. The last problem of this kind that I had to tackle actually 'tore along the dotted line of stitches' on the way back from 'the sailmaker'.

## Sewing Machines

We have a 12 volt DC electric sewing machine on board and it has certainly earned its keep. The electric motor comes close to being essential. The bulky sail material needs two hands to feed it through the machine which leaves no hand to turn the handle of a manual machine. If you are to buy a sewing machine, take a scrap of sail cloth with you and ask to see how well it will perform when required to sew four layers of fabric. If it passes this test then it should be good enough for the job.

The greater the distance between the needle and the pillar that supports the arm of the sewing machine the better, at times there will be a huge amount of material to coax through this gap. Unfortunately the gap is usually of a standard size so you will have little choice in the matter, but I have seen a few machines with smaller than standard size gaps and these must be rejected.

The needle, thread and tension must be compatible. If they are not the result will be bunches of thread where a clean line of stitches should be or at it's worst, no stitches at all. The hand-book will tell you what the tension should be for various materials and how to adjust it. It pays to have a good understanding of the function of the upper and lower tensions. The purpose of the tensions is to pull the thread evenly into the material from both sides.

A Torrington No.10 needle is suitable for most sail repairs. This size of needle is also known as 130 LR/705 LR CH No.120. Carry a good supply of spare needles with you, they are only to be found in the bigger towns. Despite the anti-rust paper in which they are supplied it pays to coat your spare stock of needles with oil and store them in an airtight container.

If a needle begins to make a loud knocking sound when it is in use, it is probably blunt or bent. Don't continue to use a needle in this condition, it can damage the inner workings of the machine. If, when you look directly at the point of a needle it reflects light, it is certainly blunt. A larger needle is indicated if the needle breaks when you attempt to force it through several thicknesses of sailcloth. This does not often happen but it is worth carrying a packet or two of bigger needles.

Nylon thread 60 (metric) will serve for sailcloth up to 7 ounces and metric 40 for anything heavier.

Take half a dozen spare bobbins with you and store them with your oiled spare needles to keep them from rusting. Better still, some bobbins are being made of plastic now. Two or three spare drive belts for the electric motor are also worth carrying.

We used the roller foot for most of our sail work. It runs over the inconsistencies in the thickness of the sail with greater ease than any other foot.

Sewing machines are fairly simple pieces of machinery with very little to go wrong. Oil your machine regularly, make sure that all the screws are tight and above all keep it free of fluff. An accumulation of fluff can actually stop a machine working.

## Sail Repair

Repairing sails in the cramped confines of a small boat is extremely heavy work and for this reason alone it is usually man's work. So gentlemen, unless your wife has muscles to match yours, you had better find out how to use a sewing machine before you depart these shores.

Most sail repair work consists of restitching seams or patching to reinforce or replace torn areas and this sort of work should be within the capabilities of anyone.

To resew a length of seam, lay the sail out and set the damaged seam in place, gently stretching the sail along the length of the seam. When you are satisfied with the way the seam is lying, mark pairs of lines (strike-up marks) right across the seam. (Figure 5) Now, when the two parts of the seam are separated, it will be a simple matter to reposition them just by lining up the pairs of strike-up marks. If the next pair of strike-up marks fail to marry up as your sewing progresses along the seam, you are heading for a puckered seam. Things tend to get worse rather than better so it pays to stop at once and unpick some of the seam to allow the next strike-up marks to meet.

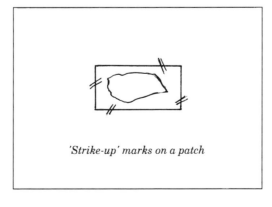

*'Strike-up' marks on a patch*

Fig. 5

Quite often when a sail suffers damage the fabric is stretched and this may require you to 'cheat a little' as you sew to ensure that the excess material is distributed throughout the length of the seam. The need to 'cheat' is even more likely to be present when you are patching a torn sail.

Ideally patches should be cloth of the same weight as the sail that is to be repaired and the warp and weft of the patch should line up with the weave of the sail. If either of these conditions are violated the sail will still drive your boat, so don't despair.

Trim out any shredded pieces of sail cloth in the damaged area and cut a patch of suitable size. Not only must the patch cover the damage but it should reach back all around so that it is sewn to sound, undamaged fabric. Place the patch in place on the sail and pencil in the strike-up marks. Sew the perimeter of the patch in position. Turn the sail over and machine the damaged edges to the patch. Don't be afraid to open a seam to allow you to slip one side of a patch neatly out of sight.

It is sometimes advocated that seams should be triple stitched for ocean sailing. There are three things that weaken the thread used for sewing a sail; the sunlight to which it is exposed, the loads imposed on it and the inescapable chafing. Three lines or two, the stitches are exposed equally to these destructive forces and it is my observation that triple stitching lasts no longer than double stitching.

Gossamer-thin sailcloth of the kind used for a ghoster can be very difficult to handle on a sewing machine. The pieces you are trying to sew together refuse to stay where you put them and any stitching you do is in danger of transforming your sail, so that it looks as if it is made of seersucker rather than sailcloth. I found that dabs of contact glue held the pieces of cloth together and gave me a reasonable chance of success.

If you are to do your own sail repairs it is obvious that you must take a supply of sail-cloth. What is perhaps not so obvious is that if you are hoping to rely on others to do your repairs you must also take some sailcloth with you, you can't expect a yachtie who does sail repairs to carry a stock of sailcloths to suit all comers' sails.

When a repair is to an area that is more complex than a straightforward seam or patch, study the work that was originally done and try to recreate what has gone before.

## Luff Rope Repair

The wire luff rope on our working jib parted when we were approaching Bonaire in the Dutch West Indies. Bonaire is a sleepy little island which didn't boast of even a make-believe sail-maker, so it was all down to us if we wanted to continue using the working jib. The luff wire was a plastic coated wire with a diameter of five sixteenths of an inch. The wire itself was only three sixteenths of an inch in diameter and this has a breaking strain about the same as a three eighth inch diameter prestretched Terylene rope. We had a stock of this size line in the rope locker so it was used to replace the luff wire. Not only did this restore the working jib to use but it was so much easier to bag the working jib now that it had a soft luff rope, that I became quite converted to the use of rope as opposed to wire for this purpose.

To renew a luff rope takes time but it is not a particularly difficult thing to do. The steps are as follows.

1. Remove the piston hanks and eyelets.
2. Note how the peak and tack eyes have been sewn in and cut both eyes free of the stitching without damaging the fabric.
3. Unpick the line of stitching that runs close to the luff rope.

4.  Cut away the little bands of stitching that go round the luff rope and hold it to the sail at about 18 inch intervals. The wire should now be free to move within the tabling (the hem on the luff). Don't pull it out yet, just make sure that it is free to move for it's whole length.

5.  Unpick a few inches of the seaming on the tabling near the break to allow you to reach the broken ends of the luff wire.

6.  Secure a length of light line to one of the broken ends with a rolling hitch and then withdraw that piece of luff rope, allowing the light line to take it's place.

7.  Attach the other end of the light line to the remaining piece of broken wire. Withdraw the wire pulling the light line through as you do so. You now have a messenger with which to pull the new luff rope through the length of the tabling.

8.  Splice one end of the new luff rope onto a plastic thimble, pull the new luff rope through and sew the thimble in place in the manner in which it was originally fixed.

9.  The next step will probably require a trip ashore. You will need to find two fixtures far enough apart to let you stretch the luff rope between them. Because the luff rope must be stretched really tightly at this stage you will need to use a block and tackle. When the rope is really well stretched between two trees, or whatever, pull the sail along the luff rope to it's fullest extent and mark the rope at the point reached by the end of the sail. This is where the second plastic thimble must be spliced in.

10. When this splice is made and secured to the sail you will find that the sail does not fit the luff rope too well. This is as it should be. The rope must be restretched and whilst it is held like this you must put in the bands of hand sewing that hold the sail to the luff rope at about 18 inch intervals.

11. Next, the line of stitches you earlier unpicked must be replaced, preferably using the semi foot (sometimes called the zipper foot) on the machine so that you can get the stitches really close to the new luff rope.

12. Repair the few inches of the tabling seam you unpicked.

13. Renew the eyelets and replace the piston hanks.

A few days after I had replaced the luff rope in our own working jib I was approached by an American yachtie who had the same problem. Unfortunately for him he had continued to use the sail for a long time after the luff rope had parted and this had caused the fabric along the whole of the luff to stretch out of shape quite badly.

The owner had no money to spare for repairs so I agreed to a sort of Lloyds form salvage operation, no cure, no pay. Payment was to be made in frozen lobsters. The result was complete satisfaction on both sides. The jib set well and so did the lobsters.

Some time later, on passage from Tongatapu towards Brisbane, on the east coast of Australia, our ageing genny exploded into a bundle of streamers. When we reached Australia we were spending money like drunken sailors on new gear, repairs and revictualling so I bought some sailcloth and set about making a new genoa, my first attempt at sail making.

## Making your own sail

We were moored in the Brisbane River on the edge of the lovely old Botanical Gardens with a major shopping centre just fifteen minutes walk away. On the other side of the river was a decaying old timber building which had been the old Royal Australian Naval Headquarters building. It was sad to see a piece of Australian history rotting away in a country that has only two hundred years of recorded history. There are Australians who care and I have a feeling that it may well be rescued but they will have to hurry.

I explored the remaining buildings and found a sound wooden floor big enough for my needs. It needed cleaning but that was no problem. When the floor was clean and dry I chalked out the full size shape of the sail complete with the necessary curves on each edge. The next day I cut the sail out. When I had finished the chalked shape of the genoa was covered with panels of fabric, each one overlapping it's neighbour by just the right amount to allow for the seams. Strike-up marks were placed across each seam and the panels were numbered so that I would know the order in which to sew them together.

The books say, 'When the cutting out is completed pick up the panels in sequence and stack them on the floor by the side of the sewing machine'. As the sewing was to be done in *Didycoy's* cabin this was out of the question, hence the numbers and marks on each panel.

The week that followed was a sewing machinist's nightmare but it finally came together. It may not be the best genoa that has ever been made but it served us from Australia to Britain and that is a lot of mileage and what's more there's a lot more miles left in it.

Should you need to try your hand at making a sail, the following points and diagrams may be of value to you.

The draught in the belly of the sail is induced by shallow convex curves at the foot and luff. When these two sides are pulled taut the sail develops its characteristic shape. (Figures 6 & 7)

The leech must have a concave curve to prevent the sail vibrating when under way. (Figure 8)

The panels of sailcloth should meet the unsupported edges of the sail either at right angles or be parallel with those edges.

In a mitre-cut sail, the single diagonal seam that reaches from the clew to the luff is called the last seam because the two halves of the sail are made and then finally sewn together at what is literally the last seam. The angle at the last seam made by the cloths from the luff must be the same as the angle made by those from the foot otherwise the pairs of cloths will not meet up at the last seam.

Probably the easiest way to decide what this angle should be is to make a simplified scale drawing of the sail on a sheet of paper. Forget the curves, just use the lengths of the luff, leech and the foot to form a triangle. Fold the foot up to lie along the leech and the fold will give you the angle at the last seam. A line drawn at 90° to the foot and another at 90° to the leech to meet at the last seam will give you the angle to be made at the last seam by the panels of cloth.

The tablings (the seams on the three edges of the sail) are made from lengths of cloth cut to a width of about five inches and folded in half lengthways before they are sewn in place.

Note how the clew and head reinforcements are arranged on your other jibs and follow that pattern.

The simplest form of clew fitting is a stainless steel triangle supported by Terylene tape.

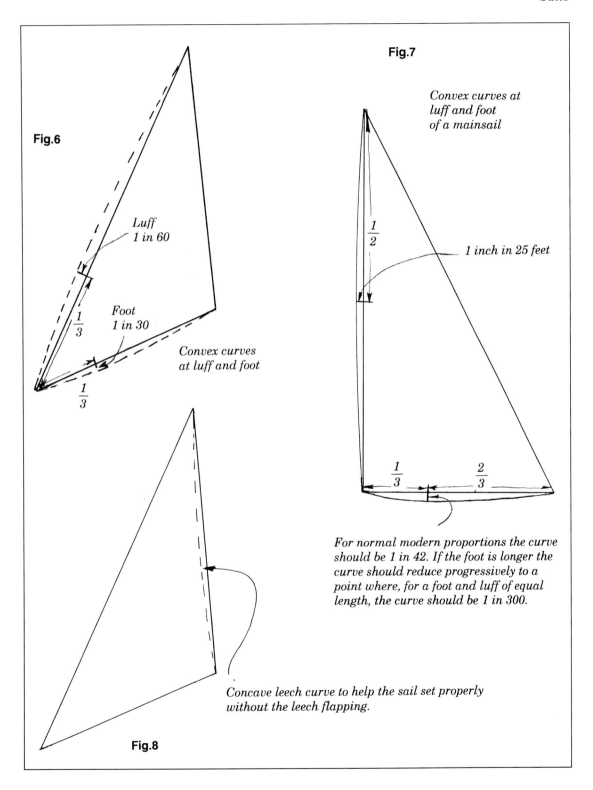

**Fig.7**

Convex curves at
luff and foot
of a mainsail

$\frac{1}{2}$

1 inch in 25 feet

**Fig.6**

Luff
1 in 60

$\frac{1}{3}$

Foot
1 in 30

$\frac{1}{3}$

Convex curves
at luff and foot

$\frac{1}{3}$

$\frac{2}{3}$

For normal modern proportions the curve
should be 1 in 42. If the foot is longer the
curve should reduce progressively to a
point where, for a foot and luff of equal
length, the curve should be 1 in 300.

Concave leech curve to help the sail set properly
without the leech flapping.

**Fig.8**

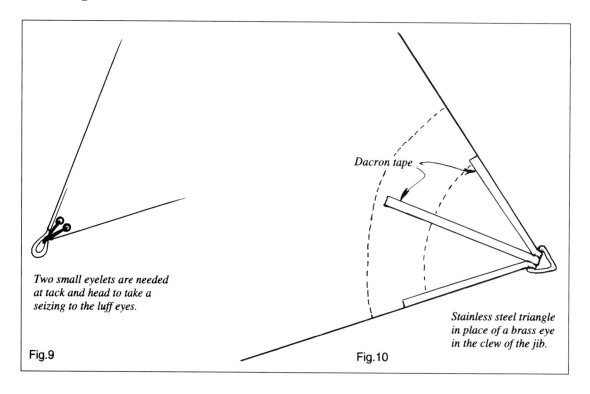

*Two small eyelets are needed at tack and head to take a seizing to the luff eyes.*

Fig.9

Dacron tape

*Stainless steel triangle in place of a brass eye in the clew of the jib.*

Fig.10

## Sail Hanks

*Didycoy* had galvanised rigging wire which needs to be oiled at regular intervals to protect it against rusting which is a chore, but on the other hand it does not part without warning as stainless steel wire is apt to. The texture of the surface of galvanised rigging wire is usually coarser than stainless and this causes bronze piston hanks to grind themselves away at an alarming rate. The hanks chafed the leading edges of the jibs and the eyelets pulled out, frequently tearing the jib as they came adrift.

We bought some tough transparent plastic of the kind that is used to cover library books. What we really wanted was some of the material that is used to make windows in boat tonneaux and spray hoods, but there was no way that we were going to find that, so we had to settle for the next best.

From then on every time I renewed an eyelet I would first cut a rectangle of plastic about 6in x 3in and fold it twice to 3in x 1½in. This was placed over the luff of the sail so that 1½in of plastic reached back from the luff on each side of the sail. The position of the eyelet hole would then be marked and punched through all four layers of plastic. Finally it would be replaced on the sail and secured there by the new eyelet.

Not only did it stop the chafe but the eyelets no longer pulled out. The piston hanks continued to wear away with the same monotonous regularity. It was not until we reached Australia and I went in search of a new supply of bronze piston hanks that I found the answer to the problem ... plastic piston hanks.

These hanks are made of the same material as the white plastic thimbles that are inserted into eye splices and seem to last for ever and a day. Once all our jibs were fitted with these plastic piston hanks I had no need to replace a single one, they lasted all the

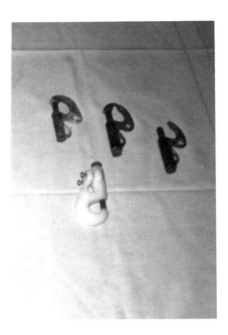

The row of three bronze piston hanks demonstrates the wear that they suffered.
The answer lies in the plastic variety shown below them.

way home from Australia and are still in use.

The design is such that eyelets and seizings are not needed, so disposing of two problems. Each piston hank is so shaped at the bottom that when it is forced over the luff rope, two small stainless steel screws hold the two sides of the plastic base tightly together with the sail sandwiched between them. As there is no movement the sail is not chafed and as a bonus they cost but a fraction of the price of bronze piston hanks. With something so good there must be a snag and of course there is. I have only ever seen them on sale in Australia.

**Headsail Roller Reefing**
We met many boats fitted with foresail roller reefing and we heard of very few failures. I suppose the ease of reefing reduces the temptation to hang onto a head sail for too long in a rising wind as sometimes happens when you have to go onto the foredeck yet again for another three rounds with a flogging jib.

Since our return to Britain I have heard that there are some varieties of jib roller reefing gear that suffer damage to the wire at the head of the gear after just one season of casual sailing. Clearly you must do some research before you spend your money. Find a large yard or marina and speak to the folk who would be responsible for removing damaged gear of this kind and sending it away for repair. They would know which brands to avoid. My feeling is that should the roller jib fail, it must be possible to lower the gear at sea. A jib that cannot be reduced in a rising wind will eventually split and the tatters will set up such a vibration that you will fear for the safety of your mast.

**Mainsail Roller Reefing**
A number of attempts have been made to produce a roller reefing gear for the mainsail. I

am not referring to the long established roller reefing gear that wraps the mainsail around the boom. The early models rolled the sail into the mast. Often they were powered by an electric motor. Anything that is electrically driven in a boat is almost certain to experience failure at some time, the small boat's environment is not conducive to the good health of electrical gear.

We met about half a dozen boats, mostly Australian, fitted with gear of this kind. They all reported problems. Most difficulties concerned the complexities within the mast, clearance and that sort of thing. In the nature of things the failure would probably not be discovered until a reef was required.

There is now a mainsail roller reefing system that is really an adaptation of the foresail roller reefing that has proved to be so successful. No attempt has been made to stow the sail in the mast thereby by-passing a number of problems. The operation of this reefing system depends on lines, blocks and a winch, all of them well tried and easily serviced items. It looks to as if it might be the right way to go.

I don't think that we met a single yachtie who depended on boom roller reefing. They had all resorted to the old style of reefing points and pendants. Boom roller reefing can be difficult to operate successfully in a really heavy blow, not that points reefing is easy at times but it does give you a chance to tame the beast step by step in a way that roller reefing does not. In really bad conditions I have spent up to 45 minutes getting our main reefed ... it's eighteen foot long boom was no help!

## Mainsails

If you are driven to attempting the construction of a mainsail you will probably find that it is a more straightforward task than is the making of a foresail. All the panels run parallel and meet the leech at 90° which is a help with sewing.

The normal head-board on a mainsail is often a work of art which is far too complex a job for an amateur to tackle. It is possible to shape and drill two $\frac{1}{4}$in aluminium plates and rivet them to each other with the well reinforced peak of the main sandwiched between them.

A slightly deeper convex curve, or roach, than is usual nowadays on the leech will let you dispense with that latter-day curse, the batten. They were introduced by our racing brethren and now one scarcely sees a mainsail without them. Before we set off we had our mainsail re-cut so that battens were not needed to maintain the shape of the mainsail. Our sailmaker promised us that it would not detract from the performance of our boat and he was right. As a consequence we never had a torn batten pocket nor did we lose or break a single batten in 40,000 miles!

## Storm Trysail

A storm trysail should be about the same area as your main when it is fully reefed. It is not usual to have a set of reefing points in a storm trysail but I think they are desirable. We spent several weeks battling our way up the northern half of the Red Sea in the wrong season and had 45 kts and more of headwinds for most of that time. Our standard size stormsail was too big for the winds we were experiencing and a set of reefing points would have been a blessing.

Modern thinking favours a taller, slimmer trysail and I must say that it makes sense. The argument is that it is less likely to be alternately filled and blanketed as the boat makes it's way from crest to trough of the very large seas that can be part of the heavy weather scene on the oceans. When this occurs the sail is filled and emptied of wind with some violence which is tough on gear and crew alike.

The tack of a trysail must be high enough to clear the furled mainsail and this is achieved by fitting a long line to the tack. The trysail can be raised on the mainsail track but if you can afford it I think a track of its own is desirable.

Modern materials are so strong that it really is not necessary to have a storm sail made of excessively heavy materials. Our sailmaker really entered into the spirit of the thing when he made a storm trysail for us. For example the entire sail has a bolt rope of 1in diameter. It would be much easier to handle if it had been taped all round with Terylene tape and I am quite sure that it would have been more than strong enough. The cloth of which the sail is made is of a weight that would do justice to the forecourse on the *Cutty Sark*. Not knowing any better at the time I assumed that he knew best and regretted my trust every time we used it.

**Storm Jib**

The storm jib is usually not more than 80 per cent of the storm trysail's area and my preference would be for rather less than that. Its foot should be cut high and rising steeply to avoid the danger of taking a heavy sea in its belly.

We had a small Yankee amongst *Didycoy*'s headsails and I decided, for reasons of economy, to use it as our storm jib and I was very glad we did. The thought of hoisting a companion to our boiler plate trysail in bad weather would have been too daunting. This sail had a luff about four feet short of the full length of the fore stay and we used a long tack line to lift it well clear of the deck. *Didycoy* seldom took any green water on her foredeck, but with the tack of the storm jib well above deck level and the very high cut clew we felt reasonably sure that we were unlikely to fill the jib with a sea that boarded us.

Cut very flat, this sail pulled like a horse despite it's size. This minute sail, on it's own, has allowed us to beat to windward making five knots and sailing within 60° or so of the wind on occasions when it would have been imprudent to raise any other sail. Despite not being made of excessively heavy fabric it has survived undamaged.

Regular and careful inspection of your sails is most important.

# Chapter 4
# Self-steering Gear

On our approach to Eua in Polynesia our self-steering broke down and we had to hand steer for six days and nights. After running sweetly for 14,000 miles, two welds had given way. So far as I am concerned you can have your 'tall ship and a star to steer her by'. Give me a snug bunk from which to listen to the gentle murmur of the self-steering gear keeping our boat on a better course than we could ever hope to steer.

When I first looked at self-steering gears I was all but overwhelmed by the sheer variety of what was on offer and it took a long time to make sense of it all. In the hope of saving you time, which becomes increasingly precious as departure day approaches, I will try to bring some sort of order to it for you.

First, for the sake of our discussion we shall divide them into three broad categories, 'Auto Pilots', 'Wind Vane Steering' and 'Self-steering', with self-steering (no capital letter) to cover all forms of steering that do not require a human hand to guide them. For our purposes we will define Auto Pilots as any gear worked by electricity and Wind Vane Steering as a gear where wind acting on a vane provides the power to turn the boat. By Self-steering (capital letter) I mean controlling a yacht's course by the use of the sheets and sails and I do not only refer to correct sail trim.

## Auto Pilots

An Auto Pilot is the only form of control that can be set to steer a compass course without reference to the wind direction. Many can also be set to sail a course that is relative to the wind direction but naturally it will then alter course if the wind changes direction.

Providing that your sails are trimmed to help the boat stay on course, and this applies to all forms of self-steering, they do not use very much power. Obviously if the boat is reluctant to stay on course because of poor sail trimming the Auto Pilot will have to work that much harder to maintain control.

Anything electrical can break down and Auto Pilots are no exception but to be fair, we have not met many people in trouble on that score. The Auto Pilot's real weakness lies in the fact that as the wind strength increases, a point is reached when it is no longer powerful enough to cope. The reverse applies to a wind vane, the stronger the wind blows the greater the power output, therefore the more efficient the self-steering gear becomes. And this bonus comes when you most need it.

Usually the Auto Pilot is harnessed directly to the tiller or wheel that operates the normal (primary) rudder but I have seen several boats where the Auto Pilot was operating an ancillary rudder. An Auto Pilot could also be harnessed to control a trim tab on either primary or ancillary rudders. The primary rudder is the normal rudder that comes with the boat. An ancillary rudder is an additional rudder that is part of the self-steering gear.

A trim tab is a simple device that utilises the power of the water flow past the rudder it serves. It can best be likened to a small rudder hung on the after edge of a rudder. In action the trim tab is turned in the opposite direction from that which the rudder is required to take. The water passing the rudder strikes the trim tab and in so doing pushes the rudder in the desired direction.

An Auto Pilot used to operate a trim tab requires less power than one that is connected directly to the rudder. This means, of course, that a smaller (cheaper) Auto Pilot can be used.

*Didycoy*'s ancillary rudder is hung well back from her primary rudder. Placing the self-steering rudder that far back gives it that much more leverage and allows you to use a smaller rudder. The cross section of our ancillary rudder is that of an aerofoil and the trim tab follows the line of that shape.

Fig.11

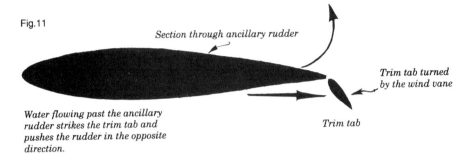

*Section through ancillary rudder*

*Trim tab turned by the wind vane*

*Trim tab*

*Water flowing past the ancillary rudder strikes the trim tab and pushes the rudder in the opposite direction.*

The trim tab on the trailing edge of our wind vane system rudder can be clearly seen here. Fig.11 shows a section through it and indicates its operating mechanism.

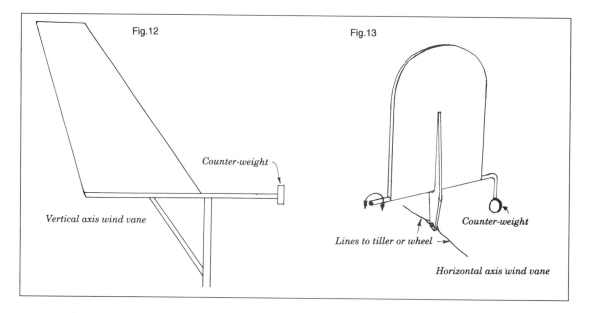

Fig.12

Fig.13

Counter-weight

Vertical axis wind vane

Counter-weight

Lines to tiller or wheel →

Horizontal axis wind vane

Fig 12 shows a vertical axis wind vane and Fig 13 the horizontal.

### Wind Vane Steering

Wind vanes come in a variety of shapes and sizes but fundamentally there are only two kinds, horizontal axis or vertical axis, Figures 12 & 13. It is claimed that the horizontal axis vane is more powerful than a similar sized vertical axis vane. Set against that, the vertical axis vane is simpler. They both do the same job and there is no basic difference in the way they are used to control the boat's course.

So long as the boat stays on course in relation to the wind, the vertical axis wind vane will present its leading edge to the wind. The moment the yacht goes off course, wind pressure will be increased on one side of the vane. This pressure will be transmitted to the steering control which in turn will bring the boat back on course. I am using the term 'steering control' because there are a number of different hook ups and we will examine them later.

The vertical axis wind vane must have a counter weight to balance the weight of the vane when the boat heels or rolls. (Figure 12). We opted for a sailcloth sleeve on a lightweight tubular steel frame. It worked quite well but the sleeve needed to be renewed after three years of tropical sunlight. The boats that had plywood vanes found that they needed to have a spare vane in the store as they could be broken.

Vertical axis vanes seem to have been subjected to all sorts of modifications in an attempt to gain additional power. All I can say is that we made our vane gear in accordance with the specifications laid down in both Bill Belcher's and John Letcher's books and have been well satisfied with the results.

When a horizontal axis vane is moved out of alignment with the wind, the wind pressure on its side tips the vane over to the opposite side. This movement is the power output that moves the steering control. (Figure 13). The axis of a horizontal wind vane must be inclined away from the horizontal by anything from 5° to 35° according to the particular design requirements. If the axis is truly horizontal the vane will either stand upright or lie

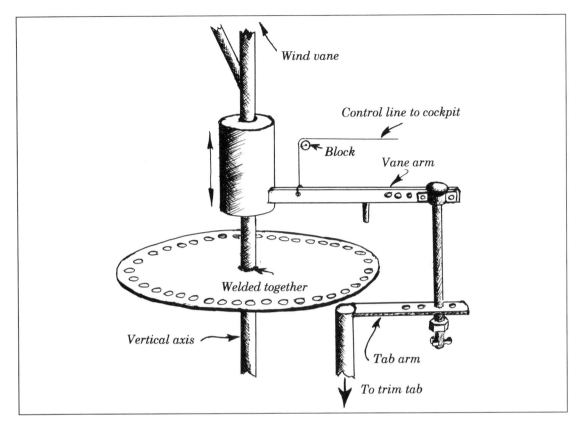

Fig 14. Wind Vane self-steering system clutch gear as used on *Didycoy*.

flat, with no shades of behaviour between the two extremes. Such behaviour will produce erratic course correction. By raising one end of the axis this all or nothing at all behaviour pattern is changed to a more acceptable form. The counterweight required by the horizontal axis vane is low slung and there primarily to make the vane stand upright.

Below the wind vane is the clutch and as its name suggests, its task is to engage the wind vane with the rudder it is to serve and it is common to both forms of vane. There are numerous versions of the clutch and many of them are illustrated in the two books I recommend later in this chapter. The point to keep in mind is that they are all intended to do the same job, but some are better suited to one set of circumstances or another. We adopted our particular form of clutch because we have a centre cockpit and it was possible to operate it from a distance. If you have an after cockpit, other forms of clutch are available to you simply because your gear is within reach.

Figure 14 is a drawing of *Didycoy*'s clutch and it will serve to illustrate what is needed, but do remember that others may look very different.

Welded to the vertical axis of the wind vane is an 8in diameter steel plate. Close to the edge of the plate are 36 holes, three eighths of an inch in diameter. Above the disc is a heavy steel collar that is free to slide up and down the vertical axis. Welded to the collar is a 1in x one eighth inch bar of steel about 17in long, with a fitting bolted to its side to take a bolt. Close by on the edge of the bar is welded a short, ¼ in diameter steel pin. When this arm, the vane arm, is raised by a pull on the control line, the clutch is disengaged. When the vane arm is lowered the pin engages with the appropriate hole in the disc, and so, if the wind vane moves the vane arm goes with it. (Figure 14) The steel collar is intended to make the vane arm rise and fall smoothly.

Below the vane arm and parallel with it is a second arm called the tab arm because it is connected directly to the trim tab. If the tab arm moves the trim tab moves and if the trim tab moves the rudder moves.

The vane arm and the tab arm are loosely connected by the bolt that lies in the fitting at the end of the vane arm. This connection must be loose to allow the vane arm to be lifted to take the pin out of the hole in the disc when you want to disengage the vane.

This type of clutch is sometimes criticised on the grounds that the limited number of holes in the disc limits the precision with which a course can be set. The argument is, that with only 36 holes to choose from you are confined to 10° increments of course. Whilst this is true, it does not mean, as is suggested, that you can be up to ten degrees off course. If you are relying on the wind vane alone to set your course you need never be more than five degrees from your chosen course.

Imagine that we wish to maintain a heading of 015°. If one hole gives us 010° its neighbour will give us 020° and that is the worst possible case, five degrees of error, not ten degrees. Any other course between 010° and 020° will be closer than five degrees. Even this five degree error need not be tolerated once you become familiar with the vane gear and the behaviour of your boat.

In action, with the gear disengaged, the yacht is put on course and the sails trimmed so that the boat holds that course to the best of its ability. When this is done the vane will be weathercocking into the wind. Engage the clutch and watch the course. If your vessel pulls to windward a little, ease the mainsheet. If, on the other hand, she wants to sail slightly downwind of the desired course, haul in a little mainsheet and that should bring her back on track. We have found that we have a surprising degree of control over the course sailed by manipulating the mainsheet in this way and the five degree limitation disappears completely.

Usually we allow our primary rudder to float free but occasionally we find that we need to give the vane gear a little help by lashing the wheel. I suspect that on these occasions I have trimmed the sails badly but it is not easy to be sure.

If you use a vertical axis vane to control a trim tab, it is important not to mount the vane directly onto the rudder head so as to give a direct drive from the vane to the trim tab. Its pure simplicity makes this a tempting solution if your boat lends itself to it, but you will be putting trouble in store for yourself. With this arrangement most boats will over-correct very badly, wandering anything up to fifteen degrees either side of the course whenever the vane attempts to bring the boat back to her proper course. A vane mounted on the boat, or more likely on a small platform that extends abaft the ancillary rudder head, will overcome the problem.

Another source of trouble can be the position of the bolt that connects the vane arm to the tab arm, it must be directly over the pintle of the rudder that is served by the trim tab. In Figure 14 you will see that both the vane arm and the tab arm have a number of holes in them to allow some adjustment of the bolt's position. If the bolt is not in the right place then your gear will either over- or under-correct. Once this ideal position has been found the additional holes become redundant.

All moving parts must be able to move very freely indeed, the least friction in any part

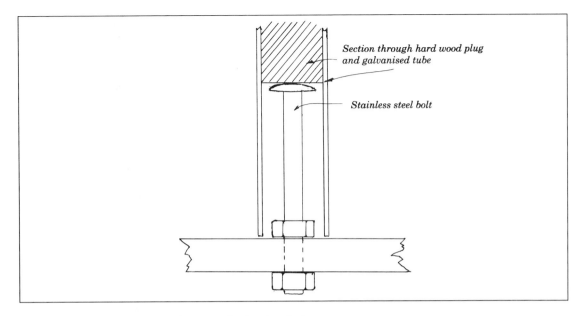

*Section through hard wood plug and galvanised tube*

*Stainless steel bolt*

Fig 15. The bottom bearing for a vertical axis wind vane.

will cause the gear to lose efficiency. It is worth installing hard plastic bushes on all the bearing surfaces and if you do, take a spare set with you. It will take a long time but eventually they will wear out. It is so much easier to get two sets made up initially than to try to remember the precise dimensions and to find someone able to do the job for you later on.

If you are making your own gear, you will need a bearing at the lower end of the vertical axis. We used the simple set-up in Figure 15. The hardwood plug had to be replaced once in forty thousand miles.

If you do not wish to use an ancillary rudder, or to link the vane to a trim tab, it is perfectly possible to use the power to turn a steering wheel or move a tiller. Lines from beneath the clutch to the wheel or tiller will be needed and although they do tend to clutter the cockpit somewhat there is no reason why they should not be used. We chose our system for its simplicity, for the power increase we gained from the trim tab and because the clutch suited a centre cockpit.

There is a wind vane system that employs a paddle or pendulum suspended vertically beneath it in the water, so that its leading edge is in line with the fore and aft line of the vessel. When the boat moves off course the pendulum's equilibrium is disturbed and it swings to one side, turning a quadrant that in turn pulls one tiller line and eases the other.

At anchor, the paddle is at risk from dinghies and mooring lines. It can be hinged up to the horizontal position which reduces the chance of damage but it does not exclude the possibility. Certainly spare paddles should be carried if you are to opt for this model. Because this type of vane steering gear is intended to move the tiller or wheel, lines will encroach on the cockpit space. At sea the pendulum type vane gear is certainly efficient and that is where it really matters.

No wind vane is very effective when running in light airs. The yacht's speed must be subtracted from the speed of the wind to obtain the relative speed of the wind passing over the wind vane, thus further reducing the power of an already inadequate wind. Beating in the same wind would mean that the relative speed of the wind would be increased.

A light wind with a confused following sea can reduce the efficiency of a vertical axis vane gear considerably. I have found that a length of line secured to the after bottom corner of the wind vane and allowed to trail in the sea astern can often improve the performance of the self-steering gear in these conditions.

A horizontal axis wind vane is handled in much the same way as a vertical axis vane. Whilst it is true that they are most usually used to operate a tiller or wheel it is possible to link them to a rudder or trim tab. I think, however, it is fair to say that the direct linkage is more difficult to achieve with this kind of vane.

If you intend to make your own vane gear to save money, we did and it has worked well. When you have digested my outline I suggest that you read two books; John Letcher's book, *Self-steering for Sailing Craft*, published by International Marine Publishing Company, Camden, Maine, U.S.A. and Bill Belcher's *Yacht Wind Vane Steering*, published by David and Charles. They are both excellent books written by people who have sailed many thousands of miles in small boats and have a real understanding of the fundamental principles of self-steering gears for yachts. They will both give you sufficient information to allow you to build your own gear, designed specifically for your own boat.

## Self-steering

There is a sheet-to-tiller form of self-steering that appeals to me immensely. We have sailed many miles with it in control and the cost was less than five pounds, perhaps ten pounds at today's prices. Sadly it requires a tiller which made it impossible to use when we changed to *Didycoy* because she is wheel steered.

If you have ever sailed a dinghy you will be aware that when the wind gusts the dinghy will try to haul up to windward. To keep the dinghy on course it is necessary to pull the tiller to windward. At the same time the load on the mainsheet is increased. The delightful thing about this sheet-to-tiller self-steering is that it uses the increased pull of the mainsheet to move the tiller to windward to counter the misbehaviour caused by the increased pull of the mainsheet. When the gust subsides the tiller is allowed to return to its normal position.

The gear we used in a 29ft Trintella consisted of a length of 10mm diameter line, a length of three eighth inch outside diameter rubber tubing and two small blocks of a size that will accept the line. As ever, friction is the great enemy so be generous with the size of the blocks you use. The rubber tubing should be of the kind used for underwater spear guns. Shock cord for some reason is a poor substitute for the rubber tubing. One block must be secured to the cockpit coaming level with the forward end of the tiller and the other to the other side of the cockpit opposite the first block.

Secure the line with a rolling hitch to the standing part of the mainsheet about two feet above it's deck fixing. Reeve the free end through the leeward block, across the cockpit, through the windward block and finally secure it to the end of the tiller. A small jamming cleat screwed to the inboard end of the tiller makes for easy adjustment of this line.

You must be sailing before the line can be secured to the tiller because you will need to have the weather helm that is normal for your boat in the prevailing conditions held on by the pull of the line from the mainsheet. To achieve this, the standing part of the mainsheet must be deflected. It's a matter of experiment to find just how much the sheet must be pulled out of its normal straight line to balance the pull on the tiller. The rubber tubing is

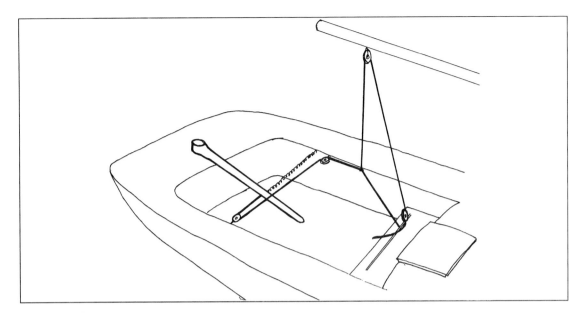

Fig 16. This simple and cheap form of self-steering is highly effective — only with a tiller.

secured to the end of the tiller and taken to the leeward side of the cockpit so that it just fails to exert a pull on the tiller when it is in the fore and aft position.

When it is working correctly your boat will stay on course as it would with the most expensive gear, whilst the tiller is pulled to windward and then allowed to return to its normal position when the wind has dropped sufficiently to let the rubber tubing assert itself.

The simplified sketch in Fig 16 shows the fundamentals of the method. Note how the deflection of the standing part of the mainsheet provides the pull to maintain the correct amount of weather helm. When the boat is on a steady course the rubber tubing is stretched just enough to balance the pull of the mainsheet. The standing part of the main sheet must be used to avoid the possibility of the thin line being dragged into a block if the mainsheet is adjusted.

If the wind pipes up, as opposed to just gusting, it may become necessary to increase the deflection of the sheet and eventually it could reach the stage where you would need to reinforce the rubber tubing with a second length to keep the two pulls on the tiller balanced. The clue to this state of affairs is the fact that the boat will be repeatedly luffing, which will not happen if the system is properly set up. It could be said that 'If she luffs in the puffs she needs stronger elastic!'

There are numerous variations on this theme provoked by the variety of sizes and designs of sail and hull, wind conditions and the point on which the boat is sailing. On a broad reach, for example, it may become necessary to change to the jib sheet for your power supply. However, it may be that the pull on the jib sheet could prove to be too strong. If this happens then a 'fool's tackle' would need to be rigged to reduce the pull.

A normal tackle is intended to supply a mechanical advantage, e.g. a pull of 10 lbs will be transformed to rather more if it is applied through a tackle. The penalty is that the rope has to be hauled further than the load is moved to gain the advantage. A 'fool's tackle' is simply a tackle that is rigged back to front so that the mechanical advantage becomes a mechanical disadvantage.

If you can set a very small jib in addition to the jib needed for the circumstances you are experiencing it is highly likely that its sheet could provide the power needed for your self-steering.

These variations and numerous others have all been used with considerable success. The principle remains; get the right amount of deflection on the sheet and balance it with the correct length and strength of rubber tubing and you will be in business. I have read somewhere of a couple who crossed the South Pacific with this form of self-steering gear but their rubber tubing perished long before they reached the far side of the Pacific. The defunct rubber tubing was exchanged for a length of line which passed through a block attached to the tiller and back to another block on the leeward side. From this line was hung a plastic bottle into which they put just the right weight of water to balance the pull of the tiller. With this system of self-steering, just as with a wind vane self-steering gear, a small adjustment of the mainsheet will bring about an alteration of course.

## Sail Balance

As was stated earlier all self-steering systems require that your boat should be tuned and the sails so adjusted that it comes very close to holding the course you want without the help of the self-steering gear. This is just one more area where you will be glad that you bought a long keeled boat. Modern racing boats seem to be designed to broach at the slightest provocation. They have become so much the norm that I am regarded with disbelief when I say that *Didycoy* has never attempted to broach in the course of 40,000 miles and all kinds of weather, but it is a fact.

For us to be able to discuss tuning there are three terms that must be understood, lee helm, weather helm and the Centre of Lateral Resistance (CLR).

A boat that heads off away from the wind when it suddenly gusts up is asking for trouble. This characteristic is called lee helm because the helm (tiller) must be put to leeward to bring the boat back on course.

Weather helm is the reverse of lee helm. A very little weather helm is acceptable, indeed it can be in the nature of an insurance. Far better to turn to windward if a heavy gust strikes your boat but best of all is the boat that simply stays on course. A boat with so much weather helm that it flies to windward at every little gust and fights to stay there in the heavier gusts is both dangerous and exhausting.

The Centre of Lateral Resistance of a boat's underwater shape is the point about which the boat will pivot when it turns. If the sail area abaft that point is too large you can see that it will push the stern away from the wind or, to state it another way, it will force the bow to turn into the wind and that is weather helm. If, on the other hand, the sail area forward of the centre of lateral resistance is too large the bow will be pushed away from the wind when it pipes up. This is lee helm and you want none of it. In tuning your boat, the ideal to aim at is to have no lee helm at all, not only does it give quite the wrong feel to a boat but in some situations it can be positively dangerous.

If your boat exhibits undue weather helm you could be to blame. There is a tendency on the part of some helmsmen to haul the mainsheet in far too hard and this will cause your boat to luff up. Ease the mainsheet until the sail just begins to show signs of fluttering, especially close to the mast, then haul the main in until the incipient flutter disappears. If the weather helm has also disappeared you will now know a little more about your boat. Easing the sheet like this can increase your speed as well as your comfort.

If the weather helm is still with you, ask yourself if the jib and the main are a proper match for the prevailing conditions. It is possible that the jib you are flying is too small for the area of mainsail you have exposed or, is it time to put a reef in the mainsail? Weather helm that won't go away with either of those treatments needs something a little more drastic.

The first thing to try is raking the mast forward. In other words slack off the backstay a little and take up on the forestays, adjusting the shrouds as necessary. If this fails, and it has to be a bad case of weather helm not to respond to this treatment, then you must move the mast forward an inch or two if the boat design makes this possible. It is surprising how great an improvement can be made by a very small movement. I remember we cured terrible weather helm in a 29ft boat by moving the mast forward two inches.

Obviously if your boat suffers from lee helm the reverse of this treatment is needed.

A healthy sea running — as seen from the shelter of the dog-house.

# Chapter 5
# Instrumentation

Like so many things in today's world of sailing the banks of electronic, go faster bolt-on goodies that adorn the majority of cockpits to be seen around our coasts, stem from racing practice. It is assumed that if it is good for racing it must be good for the rest of us and this is far from true. If you have money to spare and you enjoy playing with these bits and pieces, well and good. For those of us with tighter purse strings it is an area where immense savings can be made without an iota of lost efficiency or safety. At the same time you will be striking a blow for a return to sanity in sailing. As a reward you will develop an enhanced rapport with your boat of a kind that you can never experience if you insist on sailing by numbers.

The yachtsman who needs a meter to tell him which way the wind is blowing and that he should reef is not a fit person to go to sea. He should stick with estuary sailing until he knows better. A sailing vessel is a near living thing and to attempt to sail it as if it were a machine is a sin against nature. I am sure that anyone who sails that way deservedly misses out on a great deal of pleasure. To be at one with your boat, the sea and the weather is unforgettable with the bonus of saving a great deal of money.

Anchored off some dazzlingly beautiful beach I would often take Betty ashore in the dinghy so that she could shop at the nearby village and I could get back to the servicing and maintenance that was always waiting to be done. On her return there was the problem of attracting my attention so that I could collect her from the shore. This was a problem for most boats and quite a number had solved it by buying two CB radios with which to maintain contact. We bought a whistle, it never went wrong, needed no maintenance, the batteries never failed and it cost but a fraction of the price of two CBs. 'Over and out', as they insist on saying.


**The Log**
We had two logs in *Didycoy*. One was impeller-operated with a dial to show the speed and distance sailed. Unfortunately as soon as a barnacle decided that the impeller was a good place to roost the log would start to under-read.

We did an overnight run from Barbados to Bequia in the West Indies. There was a brisk wind and spray was flying everywhere. The doghouse windows quickly became coated with salt with a consequent reduction of visibility. With so much spray around I was reluctant to open the doghouse windows too soon so I set a figure on the log at which distance I would open the windows.

The island of St Vincent had a light with an eight mile range on its south-east corner and we were using this as our landfall. Bequia lies immediately to the South of St. Vincent. About twelve miles off St. Vincent, according to the log, I opened the windows and very nearly died of fright. Not only could I see the light with the eight mile range but every street light stood out as a separate light, we were that close. That was the last time we relied on the built-in log!

Some weeks later on passage in the open sea, we streamed our Walker log for the first time. In no time at all it and our fishing line became inseparable friends and ended in a solid ball of line hard up against our self-steering rudder. It took best part of a forenoon to sort out that one.

I know that it is traditional thinking that a log is an essential piece of gear and in the past I have preached this as strongly as anyone. Now I wonder; is it so important for a yacht on the ocean to know how far it has travelled in the past 'X' hours? Our usual navigational procedure when we were well offshore was to take a forenoon sight which would be crossed with perhaps two more later in the day. For fixes of this kind all I needed to know was the distance made good between the two sights, usually a matter of three or four hours sailing.

*Didycoy*'s maximum hull speed is about eight knots and it is pretty obvious when she is doing that, so our speed must be somewhere between zero and eight knots, both ends of the scale so obvious that there could be no doubt. One soon becomes accustomed to making about one hundred and twenty miles in twenty four hours in the trade winds, this is a steady five knots. So now we have pinned down 0, 5 and 8 knots. Surely it can't be too difficult to estimate your speed if it falls either side of five knots.

Let us suppose that there is an error of half a knot in our estimation. With three or four hours between sights this would misplace our first position line one and a half to two miles at the very most and quite probably rather less. Is this going to matter in the open ocean? After all the Pacific is somewhere between eight and nine thousand miles across!

We ceased to use a log and relied on our judgement for most purposes well before we reached the Panama Canal, rating a fish supper rather higher than the knowledge of the precise distance run between sights. If the moon was up and about during the day, as it often is, and gave us a position line that would cross with a sun sight, then we would use it to give a position without the need to use a running fix. When we were approaching a hazard and needed to sharpen up our navigation, star sights would be available at both dusk and dawn to give a simultaneous fix. Again there would be no need to know our speed or our distance run. If it became important to keep a check on our distance run we would stream the Walker log, having first retrieved the fishing line, of course.

Many of the islands we visited rose out of the water to a substantial height and could be seen from many miles off so simplifying our approach to them. Atolls and reefs were a different problem. They are so low lying that a more cautious approach is dictated. Even so I always preferred to rely on the sextant rather than the log. The sextant can give a position that takes leeway, current and drift into account. The log is incapable of measuring these important factors. The accuracy of our landfalls was always a source of pleasure to us and it was the sextant that made it possible.

In as much as most of our courses had a large westerly component a late afternoon sight, taken when the sun was dead ahead, would give a position line that was at ninety degrees to our course. In other words it would tell us how far along our course we had progressed. Approaching land a sight of this kind could be of considerable help in deciding how to manage the boat during the hours of darkness.

## Position Fixing

The sextant is a beautiful tool and the more you use it the more uses you find for it. Decca, Loran and Consol are undoubtedly wonderful instruments but as their facilities do not extend to most of the world's truly interesting places they have no value for the blue-water sailor. Transit Navigation Systems (Sat Navs) were a remarkable innovation. To pluck a position from the air and display it on a tiny screen spanked of black magic.

Fixes could be obtained at intervals that varied from about twenty minutes to six hours depending on your position in relation to the available satellites. A fluxgate compass and a suitable log could be interfaced with the Sat Nav to give DR positions between fixes just in case you were unable to work them out for yourself.

Some models were less reliable than others, one seemed to last for about 500 miles before needing the expensive attention of a skilled repair man. Whether it was a design fault or the result of mishandling or poor installation I do not know but it is a fact that as our six year circumnavigation wore on, so we met more and more yachties sitting in harbour waiting for the return of their black boxes.

It would seem sensible to ordinary mortals to send an item away for repair to ... shall we say the UK and arrange to have it returned to your next port of call. This is too simple for many of the Customs Services around the world. They insist on detaining a vessel until the gear is returned. What happens if the equipment is lost *en route* I never did discover but come to think of it, it might be the reason why some yachties dropped out of sight, never to be met up with again!

Most of our fellow passage-makers who were waiting for the return of their Sat Navs had genuflected in the direction of an alternative method of navigation by carrying a sextant. Many, of course, had the necessary knowledge and information to enable them to navigate by sextant. Others had bought the cheapest plastic sextant they could find and carried it as if it were a nautical St Christopher's medal often stowed in the vegetable locker under a sack of potatoes. Their knowledge of the sight reduction process was as poor as their sextants and I have been asked for enlightenment on a number of occasions by yachties who had discovered that there is more to successful navigation than pressing buttons.

Once astro navigation has been stripped of its mystery it only requires a few hours effort to gain a working knowledge. But what of the sextants that were so poor as to defy any attempt to take an accurate sight, no matter how much checking and correcting was done before a sight was attempted?

The Transit Sat Nav system is about to come to an end. The satellites that provided the information for the sets to give a position are falling out of orbit and will not be replaced. This means that there will be a large number of redundant Sat Navs on offer at bargain prices. Keep this in mind when you are shopping around and don't let anyone sweet talk you into buying one, unless of course you aim to set up a 'Museum of Navigation'.

## GPS - Global Positioning System

The Global Position Satellite System (GPS) has taken the place of the Sat Nav. If the Sat Nav system spanked of black magic, I think it is fair to say that the new system *is* black magic!

Brian Ash, Managing Director of Mantsbrite Marine Electronics, has been kind enough to give me several hours of his time to indoctrinate me ... there can be no other word ... on the mysteries of the system. Brian and his staff are absolute enthusiasts for all things electronic and their enthusiasm is catching.

The new system is based on twenty four new satellites, each of which emits a continuous stream of radio signals. When the signals from two or three satellites are picked up by your receiver they are turned into a position which is displayed on a small screen as a latitude and longitude ... to three or four decimal places of a minute, if you please! An incredible standard of accuracy that is far beyond the ability of any of us to plot on a chart.

Mantsbrite has a car park adjacent to its premises with room for about twenty cars. We took a small hand-held GPS into the car park. Within a minute or two of switching on it had latched onto three satellites and was displaying our position, course and speed as we walked across the car park. This information was updated every second. Quite soon we captured the attention of a fourth satellite and this gave us our second by second height above sea level, a facility for which I trust you will not find too much use! It was a truly remarkable demonstration. This, please note, is from a man who has done a circumnavigation and when asked what electronics he carried has been seen to think carefully and then to admit to carrying a large torch.

The GPS we were using measured about seven inches by three by one and a half inches. This particular model is powered by three penlight batteries which should give about 600 hours of continuous use. It's only failing that I could detect is that the battery compartment is not watertight. This matters because the set must be taken into the cockpit to operate. Being hand-held it has a built-in aerial which must be exposed to the sky to receive the signals and so it could suffer in inclement weather. However for an extra fiver you can purchase a purpose-made transparent, heavy duty plastic bag.

From this 'simple' model, the GPS ranges up to one which will do a dozen or two other things for you including reporting on the temperature of the sea, something for which we never felt the need ... but then we are not keen swimmers. A facility that does impress me is the 'Man overboard function'. On some sets, at the touch of a button, the position is recorded and a continuous distance and bearing back to that spot is given.

## Systems Integration

So far I have only written of position fixing and related services but there is more, it is called Systems Integration. It is exactly what it says. Your radar, speed and direction instruments, wind instrument, GPS, chart plotter, waypoint indicator, autopilot, teamaker and goodness knows what else, are all linked together in one integrated system. When this is done you are able to direct the instruments to conduct a multiplicity of manoeuvres without human interference and get down to the serious business of drinking pink gins and dry Martinis.

If your bank manager will allow you to indulge in such financial extravaganzas you don't need my advice, but you do need Mantsbrite's expertise or that of other similarly dedicated experts. On the face of it all that needs to be done is to connect each piece of gear to the next one and switch on. Alas no. Each item 'talks' to all the others and they must all speak precisely the same language. Something as simple as one piece of apparatus giving a position to three decimal places of a minute and another taking it to four decimal places could bring the whole expensive edifice to a complete stop.

The other area that is beyond the knowledge of most of us is that which concerns codes of practice. I'm sure that they exist now but just wait until Brussels gets to hear about the possibilities! I am told that most failures can be attributed to poor installation. Even the simplest set-up consisting of a GPS, an aerial and a power supply will benefit considerably from the attentions of a good pro when they are installed. Installation costs are high, after all you are using the services of a highly skilled operator, but it makes no sense at all to spend this sort of money and then skimp on the installation.

The complexity of the computer programmes that control the GPS can be the source of problems arising in new models. A minute error in the programme that slips past the doubtless hundreds of hours of checking that must take place before production begins could cause problems. I think the moral here is a mildly immoral one ... don't rush to buy the newest item on the market. Let someone else experience the problems first!

The belief is that the life of the GPS receivers will be a long one but it can only be a belief as none of them have yet been in use long enough to indicate how they will fare. Because the Transit Navigation System has been superseded it is reasonable to wonder if this is likely to happen with the GPS. My view is that the system is so efficient that no one is going to try to improve on it. How could a future system improve on an updated position every second with a world-wide accuracy of between 15 and 100 yards plus a statement of the course and speed over the ground? If it was possible it is not needed, so where is the profit? The answer to that question is, 'for the military'. The GPS system made possible the incredible feats of bombing during the brief war with Iraq. We all saw on TV, bombs entering ventilators on the otherwise impregnable shelters, GPS helped to make that possible.

The military version gives an accuracy of a 10 mm cube in space which boggles my mind if not yours. It occurred to someone that if this became generally available some maniac could fill a model aircraft with Semtex and direct it through the President's window in the White House. The result was that the Americans degraded the information from the satellites so that the accuracy of the sets available to the public is reduced to between 15 and 100 yards and you, and maybe me too, will have to learn to live with such problems. The military retain their 10 mm cube standard of accuracy.

I was given numerous examples of folk who'd had these systems fitted and had then immediately set off on a long cruise. This is most unwise. With the simplest of GPS the experienced navigator could manage with perhaps thirty minutes instruction. An integrated system could require several days instruction at sea. By all means install all or some of these electronics if you can afford it and it pleases you, but do not imagine for one moment that buying the gear makes you an instant navigator: there is no substitute for learning and experience.

I was interested to learn that Brian's feelings on the question of back-up and navigational understanding coincided exactly with mine. You must equip yourself with a good sextant, anything less will be a waste of money, and you must learn to use it, and practise with it regularly, so that when your electrics give up the unequal struggle you will be able to continue to navigate. If you are reluctant to believe that your electronics could fail, ask yourself why the manufacturer proudly gives you a list of companies around the world that can repair them for you.

## Navigation Equipment

In addition to a sextant you must have a current Nautical Almanac, a means of working your sight reduction (which could range from a set of log/trig tables to a pocket-sized computer) and a watch that keeps accurate time. The watch need not be an expensive one, the cheap quartz crystal watches are wonderfully consistent time-keepers. We paid eight pounds in 1983 for a quartz crystal digital watch which lost one second every four days and still does, nine years and one battery later.

It is consistency that you want in a chronometer. Just so long as it gains or loses at the same rate day in and day out it will serve your purpose. If this rate is known and logged it will always be possible to calculate the error if you can't pick up a radio time check. Quartz watches come with either an analogue or a digital display. The digital display is the form most suited to our needs.

Having spent good money and effort on an effective back up for your GPS the thought might occur to you that the magic box is surplus to your requirements. The money saved would go a long way towards covering the cost of a small freezer which I would consider to be of much greater value to the long distance sailor.

I am bound to say that an echo sounder represents a most welcome improvement on standing on a bouncing foredeck heaving and hauling a cold, wet lead line. Indeed, I find that so many of my pilotage ploys depend on an echo sounder that I would miss it sorely. I rate the echo sounder so highly that we have always carried a spare.

We installed a radar detector which proved to be of value in some of the busier areas. A detector of this kind is basically a radio receiver which is designed to bleep when it picks up a signal from a radar beam. It is then possible to switch to the D.F. mode which enables you to take bearings of the source of the signal. If the bearing changes no danger of collision exists but if the bearing remains constant it indicates that you are on a collision course and you must take positive avoiding action. It is also possible to determine whether the vessel is moving towards or away from you. The idea is good and we have found it to be a great help in busy areas, especially in poor visibility. In those same areas in good visibility it has meant that we could, to some extent, relax our look-out, knowing that we would be alerted long before danger threatened. Unfortunately most, or maybe all ships shut their radars down when they are in the open sea, so the detector ceases to be of value in the less frequented areas.

## Radio

Our VHF was of only limited value, mostly of a social nature. Occasionally, far from shore we would meet a ship that would slow down and stop close to us. With the exception of some vessels in the Red Sea this was always done with consideration and understanding. On these occasions we would switch our VHF on and usually we would be called up. A pleasant fifteen minutes of conversation would follow with considerable interest being shown in what we were doing. This usually included an offer of a position check and enquiries as to our well-being and any special needs we may have had. At times the temptation to ask for fresh food was great but we resisted it. On the only occasion we have been in sufficient need to ask for assistance we were almost overwhelmed with kindness by the vessel concerned. Such is the fellowship of the sea and I find it quite touching that a mere amateur is admitted to the inner circles as it were.

The range of a VHF radio is limited to 'line of sight' and by the power of the transmitter. By line of sight I mean that the signal will not pop over an intervening hill to reach a station on the far side. The sort of set that is carried in most yachts will normally not exceed a distance of perhaps twenty miles or so.

Channels 16, 12, 11 and 10 were all we ever used, so a set with scores of channels is really not needed. Normally ships would speak to us on channel 16, after all there were no other vessels within many miles so the infringement of the rules was of no importance. I do recall one ship that stopped to speak to us one day in the South Pacific ... the first ship we had seen for two weeks ... curtly telling me to change to channel twelve.

Quite a number of yachts carried a ham radio and a suitably qualified operator which enabled them to stay in touch with others over vast distances. They were often a useful source of information for the rest of us, on conditions prevailing in places that we were likely to visit on our next leg.

An EPIRB (Emergency Position Indicator Radio Beacon) is an excellent safety device. As its name suggests it will send out a signal that will alert rescue services ashore and help them pin-point your position. This is a system with world-wide capability. With the help of

49

ham radio operators and nearby vessels some remarkable rescues have been achieved. Considering their cost and their efficiency when they are needed, EPIRBS are a really cheap item.

As you leave Europe or North America so Radio Direction Finding facilities become fewer and further between, not to mention less reliable. We used our set twice in six years and so I would say that this is one area where you can economise.

The minimum essential requirement for a radio receiver, is one that will operate on the 2.5 Mhz to 20 Mhz band. A set that will operate on this wave band will pick up WWV, the American station that broadcasts world-wide time signals, twenty four hours a day on 2.5, 5, 10, 15, and 20 Mhz. You will need these time checks if you are to navigate by sextant or if you are to carry a sextant as a back-up. The same set will pick up the B.B.C. World Service so giving another source of time signals as well as some news and entertainment. British Consulates usually have copies of the World Service programmes for their area and the wave lengths on which they are broadcast which they will hand out on request.

To attempt to indicate suitable radio sets would be pointless in this age of rapid electronic development. The important thing is to know what the set must do for you and not to settle for anything less. We thought that we would save money by waiting until we reached a duty free port in Gran Canaria to purchase our radio receiver. Without exception we found the salesmen in all the radio shops to be so aggressive and unhelpful that we wished that we had bought our radio VAT free at home where the assistants are usually both helpful and knowledgeable. It would have been no more expensive and a much pleasanter experience.

A cassette player and a library of tapes gave us much pleasure. If your tape deck will also record the occasional tape for you to send back to family and friends at home it will give them pleasure too.

## Power Supply

Dry cell batteries do not last long in electrical gear and can be hard to come by in far off places, so it makes sense to run all possible electrical bits and pieces off the ship's batteries. There is precious little sense in running something the size of a bus engine simply to turn an alternator to charge the batteries, the wear and tear on the engine alone is enough to justify the purchase of a small generator. Diesel engines prefer to run under load and using an engine just to rotate your alternator does not qualify as 'running under load'. What is more, the time will come when your engine will refuse to work and then you will bless the day you bought your ancillary generator.

Tropical areas have an abundance of both sunshine and wind, either of them can be used to generate power. Solar panels would be my first choice, they are efficient, they have no moving parts to wear or break and they make no noise. The old circular solar cells have given way to rectangular cells and this allows more of them to be packed into a given size of panel thereby increasing the output per panel. One or two panels of this kind can probably take care of most reasonable demands in the tropics.

For maximum efficiency a solar panel must be kept clean and unshaded and ideally at 90 degrees to the sun's rays. This last requirement is obviously a counsel of perfection that will never be achieved in a moving boat but the considerable numbers we saw in use did not suffer unduly on this score. As for reliability, it is well proven. The buoyage around the tropical coasts of Australia, for example, is powered exclusively with small solar panels.

Wind-driven generators probably have the edge on solar cells in our climate but once a boat reaches the areas of constant sunshine the solar cells really come into their own.

Whilst the wind-driven generators can produce electricity day and night they can be very noisy. With a twenty to twenty five knot trade wind blowing some models were extremely noisy. Constantly moving parts in any mechanical device have a finite life and no matter how long it may be they will come to an end some day.

A towed generator will be just as eager to make friends with your fishing line as will your log line. The few people we met with towed generators were not excited about their performance and of course they don't work when you are at anchor. It is often not realised that in the course of the average circumnavigation much more time is spent at anchor enjoying the fruits of your labours than is spent battling the elements to get there.

If the clutch on your engine is disengaged whilst you are sailing, the propeller and its shaft will auto-rotate. If you wish, it is possible to run a generator from this rotating propshaft but only while you are moving. *Didycoy*'s prop shaft rotates in this way if the clutch is left in neutral whilst we are sailing, it makes a noise as some of the elements in the gearbox rotate in harmony with the rotating shaft. The continual movement within the gearbox must be causing a measure of wear and tear and we find the constant noise is irritating, so *Didycoy* sails with her clutch engaged to stop to this auto-rotation.

As if this was not enough the autorotation of the propeller slows the boat down. You have probably heard as many inconclusive arguments on this subject as I have but it was settled for me, once and for all, by a helicopter pilot who was also a sailor. It seems that when a helicopter suffers engine failure the machine falls from the sky. If the pilot is sufficiently skilled he can throw the engine out of gear and cause the blades to auto-rotate. This so reduces the speed of the helicopter's descent that it can land with a degree of safety. Although the movement is vertical in the one case and horizontal in the other, the action and the result are the same in each case.

The small four-stroke petrol-driven generators are very good, using very little fuel and working quietly. They have the added advantage that they can be switched to producing 240 volt AC which will allow you to run power tools when the need arises. Petrol is, of course, a dangerous commodity to carry inside the yacht. Since its vapour is heavier than air it will linger in the bilge waiting to be ignited by the slightest spark. Because of the speed at which petrol ignites, ignition in this case equals explosion. If your choice is to be a petrol-driven generator you must be prepared to store the fuel in a proper purpose-made container on deck so that any spillage will find its way safely over the side.

# Chapter 6
# Electricity

Before we worked on the electrical system in *Didycoy*, our 36 horse-power engine and 50 amp alternator would produce no more than a two amp charge. With a battery capacity of 250 amp hours, a two amp charge would take a theoretical 125 hours to charge the batteries from scratch; more than five days! When the charging circuit was investigated it was quickly discovered that all the wires were either joined with a screw connector block or simply twisted together and taped. Many of the wires that should have been continuous were made up of several pieces. To make matters worse every joint was well and truly corroded. Each one of these points was resisting the easy passage of electricity which dissipated power as heat rather in the manner of an electric fire and was certainly to be blamed for at least a large part of the voltage drop. The danger of an electrical fire I will leave to your imagination.

## Circuits

At the time we could not afford to rewire the circuit so all the offending joints were parted, cleaned up and then soldered together. This resulted in an immediate improvement, the charge rate went up from two amps to fifteen amps and this is the way things were when we left Falmouth. Pleased but not satisfied can best describe my feelings at this point and as a result I talked about the problem with anyone whom I thought might have knowledge enough to inform me.

It seems that a battery quickly convinces your control gear that it is well on the way to being charged and calls for a reduction in the charge being delivered to it. A battery can happily accept a charge that is equal to 10 per cent of its amp hour capacity, in the case of our 250 amp hour battery that would be 25 amps. I was strongly urged to install a circuit that could be made up quite cheaply that would allow me to override the inadequacies of our charging system. By setting the controls in a particular fashion, I was told, it would be possible for me to choose the rate at which the batteries would be charged.

Fortunately we were in close contact with an American boat at the time. This yacht was exceedingly well stocked with electrical and electronic aids and comforts. Sat navs, fridge, freezer, various radios, weather fax, ice-maker and a second ice-maker to produce ice blocks from soda water to avoid contaminating their whisky with plain water, you name it and they had at least one and probably a spare. The owner had installed a circuit of the kind I had been advised to adopt. Unfortunately for him but fortunately for me, his over-ride circuit went berserk and severely damaged many of his electrical goodies. Needless to say the idea lost a lot of it's appeal for me.

Soon after this incident I met an American yachtie who was an electrical engineer. As a result of conversations with him we completely rewired the charging circuit. The wire that was pulled out had a cross sectional area of 12 square millimetres. The new wire has a cross section of 50.5 square millimetres. Now, if the batteries are low, the initial charge rate is 50 amps, dropping to 25 amps until the batteries are well on the way to being charged and then dropping quite low or even to a nil charge.

At the same time another change was made. We had a blocking diode in the original circuit, the function of which was to judge which battery needed how much power and meter it to the appropriate battery. It worked well enough but it was obvious that like all electrical gadgets it could fail. If it did so, you could be sure that it would happen miles from anywhere. In place of the blocking diode we fitted a large, four position switch which can be turned to 'OFF', 'BATTERY 1', 'BATTERY 2' or 'BOTH BATTERIES'. Now we can charge or use either or both batteries as we wish. In the event of an electrical fire we can switch to OFF and isolate the batteries thus removing the cause of the fire.

The switch we have installed is of a kind that industry uses. It is robust and can be taken apart to allow you to clean the points should they become corroded. It would seem that this is the only likely failure. The switches on offer to yachtsmen are usually sealed units which makes it impossible to service them in this way.

Alternators usually have a built-in regulator to ensure that batteries are not overcharged and this shows itself in the behaviour of the ammeter which will indicate a steadily dropping charge rate. If this regulator is damaged the alternator may continue to charge your batteries but the rate will not diminish as time goes on. With a maximum charge being pushed into your batteries they will soon become fully charged and start to overheat. If you have failed to notice that your ammeter is showing a continuous full charge, your first warning will be a smell of boiling sulphuric acid. If this is ignored the next step will be the mother and father of a bang and a flash as your battery explodes.

A voltmeter can be a helpful check on the state of your batteries. If a battery's reading drops below 12½ volts you can be sure that the time to recharge is approaching, a check with a hydrometer will confirm this. If, when you are charging, you observe that the amps to a particular battery are low and the voltage is high, perhaps as much as fifteen volts, that battery is probably failing. If after a good charge that battery's voltage drops fairly quickly you can be sure that it is on its way out.

Electrical circuit testers are fine if you have the knowledge to take advantage of the testing facilities they offer. If you lack these skills and your knowledge is confined to straightforward questions such as 'does the power get as far as this point?' then a simple testing lamp will serve and be very much cheaper. A twelve volt bulb, a simple bulb holder, some insulated wire and a crocodile clip is all that is needed. Make the neutral wire quite long and then you will be able to attach it to a point you know to be satisfactory some distance from the place at which you are working. If the positive wire is stiff, it will serve to prod its way into places without the need to take them apart.

Once you have a reasonable set-up in your vessel most electrical faults will either be a break in the line or chafed insulation that allows the wire to short out and blow a fuse. The correction of either of these problems requires nothing more than electrical commonsense and their diagnosis certainly does not call for sophisticated testers.

## Cables

Two kinds of cable are available, single strand and multi-strand. The vibration set up by sailing or motoring will, given time, work-harden the single strand wire and cause it to break sooner than the multi-strand variety. If you use wire that is thicker than is needed for that part of your circuit there is no harm done, except to your bank balance. A wire that has too small a gauge on the other hand will cause voltage drop and if it is extreme it could cause overheating as the power struggles to get to the far end of the wire.

The table on page 54 will help you determine the gauge of wire required for a particular situation. To enter the table you will need to know the voltage drop factor. To calculate this you must know the length of the cable run in metres and the maximum load that will be imposed on it, expressed as amps.

The cable length must be doubled as there are two wires, positive and negative. To find the load add together the wattage of each item that is to be served by the circuit and divide the total by the voltage to convert it to amps. e.g. 60 watts in a 12 volt system is 5 amps but 60 watts in a 24 volt system will be 2.5 amps.

Voltage drop should not be allowed to exceed $7\frac{1}{2}$ per cent of the voltage being used. $7\frac{1}{2}$ per cent of 12 volts is 0.9 volt. In a 24 volt circuit it would be 1.8 volts.

To find the voltage drop factor:-

$$\frac{0.9\text{volts (this is the permitted max. voltage drop)}}{\text{cable run in metres x maximum load in amps.}}$$

This is for a 12 volt system. If you wish to use it for a different voltage then you must change the 0.9 volts for the appropriate figure.

Now we can repeat the example with some figures. Let us suppose that we wish to run a cable of 14 metres in length and the maximum load will be 4 amps at 12 volts. The cable length must be doubled to take account of the pair of wires so that becomes 28 metres.

$$\frac{0.9 \text{ volts}}{28 \text{ metres x 4 amps}} = 0.008 \text{ voltage drop factor.}$$

When the table is entered with a factor of 0.008 we find that a cable cross section of $5\frac{1}{2}$ square mm would serve (diameter of about 1.8mm). When making your purchase you will probably be limited in choice so obviously it will be a case of bigger is better but not too much bigger, that would be a waste of money.

| Voltage Drop Factor | Cable x Section | Approx Dia. of Wire |
|---|---|---|
| 0.0012 | 35 sq mm | 11 mm |
| 0.0014 | 30 | 9.5 |
| 0.0017 | 25 | 8 |
| 0.0020 | 20 | 6.4 |
| 0.0028 | 15 | 4.8 |
| 0.0037 | 12 | 3.8 |
| 0.0046 | 10 | 3.2 |
| 0.0055 | 8 | 2.5 |
| 0.0072 | 6 | 2 |
| 0.0100 | 4 | 1.3 |
| 0.0180 | 2.5 | 0.8 |
| 0.0260 | 2 | 0.6 |
| 0.0310 | 1.5 | 0.48 |
| 0.0430 | 1 | 0.32 |

Sometimes it is necessary to run a cable up a stanchion or perhaps the backstay which is often used as an aerial. I have yet to find a tape that does not come adrift after a very limited period and there you are with a set of Irish pendants that get longer and longer. Finally in desperation I tried holding the cables in place with a length of French whipping; it works and it looks good too. If it is given a coat of paint it will last forever. In case you are not familiar with the French whipping, it is simply a series of half hitches.

You must take a reasonable supply of cartridge fuses and bulbs for your various pieces of electrical equipment, they can be hard to find in some places. Our alternator ceased to function when we were in the Red Sea. No problem, I had brought a spare. The spare was a different pattern and necessitated a measure of rather precise metal work in very rough weather conditions before I could mount it. Take heed, do take a spare alternator but do also make sure that it is identical with the one in service.

It is becoming increasingly common to find boats wired to accept shore-side power. You will find it of only limited use to you if you are bent on sailing to far off places, but I feel I cannot leave the subject without a word of warning.

Shore-side power plus boat plus water is a lethal combination. If it is your intention to install a circuit to accept 240 volts then be sure to incorporate a device called a Ground Fault Interrupter. The name varies a little but a good electrical equipment shop will recognise it by this title. Its function is to measure the incoming current and compare it with the outgoing current. If the comparison does not satisfy it it will instantly shut off the power supply thereby saving you from electrocution.

There are now thirteen amp plugs for domestic use that function in this fashion. I can visualise a thirteen amp wander lead socket at the inboard end of a line bringing power aboard with a cut-out plug of this kind as the first onboard fitting that accepts the power for distribution around the boat.

Many Polynesian islands have great cumulus rain clouds sitting over them.

Chapter 7
# Engines

I have to confess to a love hate relationship with our engine with hate easily outweighing the love. On the very few times it got us out of trouble it was indispensable. The rest of the time I would just as soon have given up the space to a sack of potatoes. Clearly it is not possible to sail great distances on a yacht's engine alone. One soon becomes sail orientated to the extent that the engine is seldom called into use simply to 'get in' as it so often is in home waters at the behest of the Great God Work. You become quite philosophical about the length of a passage, if good use has been made of the prevailing conditions that's all that seems to matter.

One soon learns that the average sailing boat's engine is nothing more than a fair weather aid. Our thirty six horse power diesel serves a large three-bladed prop and in quiet conditions we cruise at about 5 knots. It is possible to push the speed up to about 7 knots but the consumption of fuel goes up accordingly. In very strong head winds a throttle setting that would give us 6 or 7 knots in gentler conditions can often do no better than push *Didycoy* along at one or two knots. In the same wind a storm jib and trysail or deep reefed main would give us a good 5 or 6 knots.

If the wind should turn and pipe up when you are at anchor, putting you on a lee shore, it is your sails that will provide the pulling power to get you safely away, not your engine. By all means throw your engine's power into the fray but do raise your sails first. In quieter circumstances the engine can sometimes get you out of difficulties that sails alone could not rectify. Certainly our voyage would have ended on the San Blas reef if our engine had not got us out of that spot of trouble. That was its most dramatic contribution to our safety but it also took us out of the path of three or perhaps four ships when we were to all intents and purposes, becalmed and apparently unseen. The real virtue of an engine lies in this aspect of passage making. The bonus that comes with an engine is that it will make some interesting and delightful places accessible to you. Places with tree lined entrances for example or narrow twisting channels that will lead you into the enchanting seclusion of many tropical lagoons would be denied you if you were without an engine.

## Engine Alarms
An audible alarm that is activated if the oil pressure falls to dangerous levels or the temperature rises above a reasonable point is well worth installing. They are set at levels that give you adequate time to get to the controls if trouble is brewing. It is prudent to fit an alarm of this kind so that the buzzer is activated when you switch on prior to starting the engine. You are then assured every time you use the engine that the alarm is in working order.

## Filters
Diesel engines are used all over the world so items like filters were usually obtainable in all but the smallest places but the reference numbers were not always readily recognized. We kept a set of old filters that had been washed in kerosene to take with us when we were seeking new filters. It eased the reference number problem and also helped to overcome the language barrier in a few places.

Fuel in a limited number of places can be dirtier than you would expect in Europe and it pays to change filters on time. It may be that filters will not be available and you are desperate to change those you have in use. If you are faced with that problem it is possible to wash an oil or a diesel filter in several changes of kerosene or diesel until the fluid you are using comes away clean. This is obviously a less than desirable practice but it can get you out of trouble on the odd occasion.

## Engine Maintenance, Checks and Spares

Our engine cooling water intake seacock had a bronze strainer that rotted away part way across the South Pacific. I purloined my wife's nylon pan scrubber, cut it to size and stuffed it in place. That was eight years ago and it is still doing sterling service. I recall a fellow yachtie repairing an outboard engine. A small steel flap valve had corroded away. He replaced it most effectively with the end of a feeler gauge blade.

The hose clips you have on board are too small for the job in hand? Take two of them, open both and feed the end of one into the opposite end of the other .... you now have one hose clip of twice the length.

One of the more valuable comments ever made to me was 'every effect has a cause'. It sounds trite perhaps but stated as an easily remembered phrase it did much to make me aware of the onset of a problem rather than waiting for a more spectacular manifestation. No longer did I wipe up the odd spot of oil without asking myself why it had appeared.

Nuts and bolts should be finger tightened before applying a spanner. That way you will strip far fewer threads, your fingers just do not have enough strength to damage the thread. Shifting spanners are useful tools but they can tempt you to use a large spanner on a small nut simply because it is possible to close the jaws to grip the nut. The excess leverage created by an overlong handle makes it fatally easy to apply too much power and strip a thread or snap a bolt.

A banjo on a fuel line is held down by a hollow bolt and it is not always possible to find a replacement for the one you have just dropped in the bilge. If you can find a suitable bolt file a flat on one side of it and it will function as well as a hollow bolt.

We have always made a point of carrying a packet of Blu-Tac™. It will hold a screw on the end of a screwdriver when you can't get both hands to the job. A blob of Blu-Tac on the end of a cane will pick up a small item from an otherwise inaccessible spot. Blu-Tac is one of those materials that have innumerable uses that you cannot think of until the need arises. No boat should be without.

A 12 volt soldering iron will earn its keep. If you can't find one, an old fashioned soldering iron that can be heated on the top of the gas cooker will serve.

You can invest a small fortune in spares and still not have the one you need but it does make sense to carry some spare parts, particularly the smaller and therefore cheaper items.

Some of the small spares we had .... or would have been glad to have, were:—

> Spare gas supply line.
> The correct packing for the sea toilet pumps.
> Stern gland packing material.
> Rubber gasket for the pressure cooker.
> Grease cups .... they are so easy to drop in the bilge.
> Engine drive belts.
> Set of gaskets and seals for the engine.

Set of banjos and the plastic tubing that takes the excess fuel
from the injectors.
Alternator, same pattern as the existing one.
Split pins.
Bilge pump spares.
Light bulbs from torch to masthead.
Fuses for all the electrical and electronic gear you carry.
Brass or stainless steel eyelets, punch and fixing tool.
Impellers for all water pumps.
An engine workshop manual with exploded drawings of each area of
the engine. Even though you may be no engineer, the professional.
you engage may well find it a help and that means a saving in
time and therefore money.
Manuals for all electrical gear.

My rapport with engines is, to say
the least - tenuous!

# Chapter 8
# Anchors and Anchoring

If I lay great emphasis on anchors and anchoring it is for two very good reasons. The average passage maker's boat is anchored in 98 per cent of the places in which it stops. Poor ground tackle and poor anchoring technique probably destroys more boats than bad weather ever does.

There is a variety of anchors on offer but I think you would be well advised to stick to those that have a history of success. This narrows the choice down to a Danforth, a CQR or a Bruce. We carried four anchors. One, a 65lb fisherman's anchor, came with the boat and so it stayed with us but I would not have bought one from choice. In addition to our 65lb fisherman's anchor we carried a 35lb Bruce and two CQRs, one of 35lbs and another of 40lbs. Perhaps four anchors sounds excessive but it has to be borne in mind that it is always possible to lose an anchor and chandlers are few and far between in remote places.

The holding power of the fisherman's anchor, weight for weight, is no more than about a third of that of the others I have mentioned. We never used ours, partly because a 65lb anchor is a brute of a thing to manage in a thirty six foot boat but also because it could be difficult to retrieve if it jammed a fluke in a rocky crevice. The fisherman is often recommended for anchoring over rocky ground where it is claimed that it will have a better chance of finding a spot amongst the rocks where it will be able to hold in a way that other anchors could not. This clearly must be so but I would far rather move on until I could find better holding ground in which to plant my anchor.

The Danforth's sole advantage over the CQR and the Bruce is the fact that it can be stowed flat. It is usual to carry your main anchor on the stemhead roller and so a close stowing ability is not an important feature of your bower anchor.

Our Bruce anchor is on the left, the CQR in the middle and our brute of a fisherman on the right.

We have used our Bruce anchor every time we've anchored and have nothing but praise for it. In 40,000 miles we dragged once and that was my fault. I chose to anchor over rock covered with a thin layer of sand which gave the anchor no chance to do what it does so well, namely, dig in deeper and deeper as the load comes onto it. We found that given a chance the Bruce dug in easily, held well no matter how hard the wind blew and caused no problems when it was being raised. I also like the fact that the Bruce has no moving parts.

Having said that, the trouble with passing judgement on a particular type of anchor is that there is much more to be taken into account than the anchor alone. It is not possible to separate the anchor's performance from the length and weight of the chain used, the kind of bottom, your technique for setting the anchor in the ground and so on.

The Bruce was originally designed for anchoring oil rigs and it was only later that it was produced in sizes suitable for yacht use. The plough or CQR and the Danforth were both designed about fifty years before the Bruce and like it, they are anchors that will dig themselves in deeper and deeper as the load increases if they are laid correctly. It is this behaviour that, weight for weight, gives them about three times the holding power of the old fisherman's anchor that stops digging in once one fluke and arm are buried up to the shank.

The weight of an anchor must bear some relation to the size and weight of the boat it is expected to care for and manufacturers issue tables to guide you in your choice. It is generally reckoned that an anchor of less than twenty five pounds will not cut its way through a layer of grass or weed on the sea bed so I would say that should be the minimum weight regardless of the size of your vessel.

Any anchor must be pulled horizontally across the sea bed to allow it to dig in. The Bruce, CQR and the Danforth are so constructed that a horizontal pull will cause them to dig in and to continue to bury themselves until the resistance created by their shape and the chain that follows them down into the sea bed is equal to the load imposed by the boat. The action is not unlike that of a kite, get the string attached to a kite so that it is held at the right angle to the wind and it will soar up and up. It is this kiting action that makes these anchors easy to recover. Once the pull is changed from horizontal to more nearly vertical the anchor 'kites' it's way upwards through the ground to the surface.

## Anchor Cable

Your choice of anchor cable lies between chain, a short length of chain and the rest rope or all rope. An all rope anchor cable is a nonsense. Sooner or later it is going to chafe through leaving your anchor on the sea bed and your boat off to sea all by itself. A short length of chain backed up by rope is not much better. True, the two or three fathoms of chain will go some way to protecting your warp from chafing on the coral heads that some anchorages have scattered around, but it may not be enough to give you complete protection.

In Flying Fish Cove at Christmas Island it took us about two hours to untangle our cable from the coral heads it had made friends with. If we had been using rope instead of chain it would have chafed through in no time at all. We have been anchored in places with a good sandy bottom but when we've recovered our anchor the chain has come up bright and shiny in places where it has dragged across coral clumps as the boat has ranged about to the pull of the wind.

Whilst it is true that the short length of chain will improve the catenary of your anchor rode it really is suited only to use with a kedge anchor. To get and maintain that essential horizontal pull on your anchor with rope you will need to use at least twice the length that you would if you were using chain.

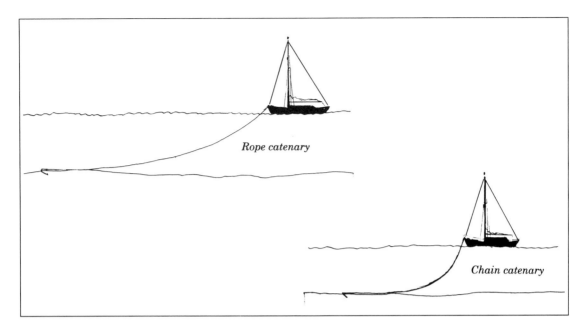

*Rope catenary*

*Chain catenary*

Fig 17. Catenary of rope and chain anchor cables.

In the course of our circumnavigation we saved three boats from almost certain loss; not one of these operations was the least bit heroic I hasten to say. In each case the boats were left unoccupied and were anchored with a small anchor and a length of rope. Each one dragged its anchor in quite moderate weather. One was floating down the Brisbane River under the influence of a three knot current on its way out to sea; another, in the West Indies, was headed for a reef and the third was on its way westwards into the Pacific and would unquestionably have been lost had it not been for our action.

How people can go off and leave their home so precariously anchored I cannot for the life of me understand. I know that in two cases it made no difference to the skipper's attitude to anchoring. In one case when I mildly commented on the relative merits of rope and chain the young owner said, 'You can't expect me to buy chain with which to anchor, I have just spent £1,000 on a sat nav'!

When you are anchoring with chain you will need a minimum of three times the depth of water at high water. With rope this must be increased to six times the depth. If it comes on to blow the scope may need to be increased to give greater security. Whenever possible we have used a six to one scope in heavy chain. By 'whenever possible', I mean in an uncrowded anchorage where we were able to swing without being a nuisance to any neighbours we may have had. Most of the anchorages you will use will be virtually tideless and you will be wind rode and not swinging to the tide as you would in home waters, so a good scope will not usually create problems for others. The chances are that they will be riding to a similar scope anyway.

If the wind strengthens the additional loading will cause your boat to sit back on her anchor and lift some of the cable off the sea bed. The effect will be much more marked with rope than it will with chain because of the light-weight nature of rope compared to chain. At its worst this change of catenary will cause the shank of your anchor to lift and start the anchor rising through the sea bed. As is probable in a rising wind, the sea will pick up and cause the boat to jerk back on her cable as she rises to each wave, adding to the strains imposed on your anchor.

Each time a boat rises to a sea and snatches at its chain the noise it makes is awful. Every snub is announced by a resounding bang that echos through the boat making sleep quite impossible. The noise can be excluded and the movement dampened down considerably by rigging a pendant of stout nylon rope to relieve the chain of the strain. We have kept about 2½ fathoms of strong nylon line stowed close to the anchor winch. This line is partly encased in a tough plastic tube about two feet long to protect it from chafe where it meets the stem head roller. In use the inboard end of this line is secured to the anchor winch and the outboard end is made fast to the chain with a chain hook. The plastic tube is then positioned over the stem head roller to protect the line from chafe.

When all is ready the chain is eased until the nylon rope is taking the load instead of the chain, leaving a shallow bight of chain hanging free. The elasticity of the nylon line absorbs much of the shock, peace descends and the anchor is relieved of much of the strain as the nylon cushions each snub. In the unlikely event of the nylon parting the anchor chain will take up the load.

## Anchoring in Tandem

There is a technique called anchoring in tandem that can be used if you have advanced warning of extreme winds coming your way. To achieve this a second anchor on a short length of chain is shackled to your anchor cable a suitable distance from the main anchor and both anchors are used as one. Not only do you gain the added security of a second anchor but the technique adds some extra weight to your cable which will help keep your normal bower anchor in place.

Our normal method of anchoring has kept us safe in winds of up to about force nine. As we have probably not experienced winds of greater strength than that whilst at anchor, I cannot write from experience but I would say that all the indications are that we could have withstood stronger winds without dragging. Nevertheless, if I had advanced warning that I was about to undergo a period of excessively strong winds I would be tempted to resort to tandem anchoring.

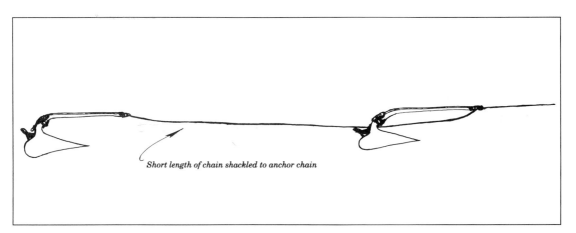

*Short length of chain shackled to anchor chain*

Fig 18. Anchoring in tandem.

## Anchor Winches and Windlasses

In a boat of more than 25 feet length overall a winch becomes a necessity. An electric winch would be excellent just so long as it can be worked by hand when the electrics fail. Ours is an old Simpson Lawrence hand-operated, two-speed winch which will take both chain and rope. A two-speed winch which can take rope as well as chain can be a blessing should the need arise. The extra power produced by the lower gear can be used to break out a stubborn anchor or to winch yourself off should you go aground in tideless conditions. The rope gipsy will allow you to use the anchor winch to bring a dinghy aboard with the minimum of effort, put a man up the mast or even recover a crew member who has fallen overboard and cannot climb back aboard. It will also let you kedge your boat over considerable distances with very little effort.

Some winches are designed to require 180° of movement of the winch handle on each stroke. Presumably this is intended to either give more power or to bring in more chain per stroke. To work with a winch handle that requires a movement that practically touches the deck at both ends of the stroke is both difficult to operate and tiring in the extreme. It must be possible to operate your winch from a normal standing position and the stroke should require no more than a comfortable arm's reach. The anchor lever must be strong enough to allow you to use your strength on it without it collapsing under your hands. This won't often be needed if the winch is any good but there could come a time. If it does, the frail chromium plated toys that are offered with some winches would fold up long before this point is reached. If your winch has a lever of this kind you must provide yourself with a stout length of galvanised steel pipe at least as a stand-by.

When you are choosing an anchor winch don't lose sight of the fact that your winch must fit your chain, do not let the winch dictate the size chain you use.

I have met a number of blue-water sailors making use of a particular type of winch that relies on a bicycle chain and sprocket for its working parts. They all said the same thing. The system worked well enough but two years of passage making was as long as the chain and /or the sprocket lasted. If you have a winch of this kind the message is clear, carry a set of spares. To have to wait for weeks for spare parts to arrive could be a disaster.

## Galvanised Chain

Shortly before we set out we had our anchor chain regalvanised. Six years later it again needed regalvanising, the anchor chain on a passage maker's boat has a hard life.

If you aim to get your chain regalvanised at any stage, make sure that the galvaniser has a piece of equipment called a 'shaker'. Without this your chain will be returned with many of the links stuck very firmly to their neighbours. In use, a chain in this condition is a confounded nuisance, alternately jamming in the navel pipe and jumping off the gipsy. The only way to separate these links is to hit them with a hammer which flakes off the precious zinc you have just paid so dearly to have attached to your chain.

## Chain Hooks

Farm sundries shops sell chain hooks, you see them on truck tail-boards that are supported by chains when they are in the horizontal position. These chains are designed so that the hook will slip over a vertical link but will be stopped by the next link because it lies at 90 degrees to the first one.

A hook of this kind shackled to a short length of strong line that is secured inboard is a nice back-up for your winch brake. When you are under way it can help keep your anchor

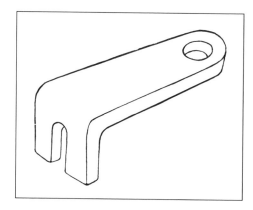

Fig 19. A chain hook.

securely in position on the bow roller. If you can't find a chain hook get a substitute made up in heavy steel shaped as in Figure 19 and have it galvanised.

## Anchoring Technique

Earlier it was stated that the technique used in anchoring is important. It might help if I describe how we go about it.

If other vessels are anchored, we approach on the same heading as they appear to be lying. If there are no other boats at anchor we head in according to the state of the wind or tide so that our heading is what we judge we shall lie at when we are anchored.

As we approach our chosen spot, my wife keeps me informed of the depth and when it is considered to be right the boat is slowed and stopped. When we are stationary I lower the anchor with enough chain out to allow it to reach the sea bed. At the same time my wife puts the engine into reverse gear and as we slowly move backwards so I pay out chain to let it lie in a line until we have three times the depth out. Then I apply the brake to the winch. If the anchor digs in under this treatment the chain lifts out of the water just ahead of the bow of our boat and then goes quite taut as it takes the strain of the reversing boat and brings it to a stop.

If the anchor holds at this point I allow more chain to run out and my wife continues to slowly reverse *Didycoy* to lay the chain out in a straight line until we have what we regard as the right amount of chain on the sea bed. The winch brake is reapplied and the engine is allowed to continue in reverse until the chain again shows that we are securely anchored. Note how the whole operation is carried out in a controlled and gentle manner. The aim is to lay the cable out in a straight line and dig it in, not to dump it in a big heap and hope that it will sort itself out later.

When *Didycoy* is settled with the engine shut off, we take two or three anchor bearings and enter them in the log so that we can check our position if we think we may be dragging. I like to find a mark on either side of the cockpit if possible, even if it is only a rock on the beach and a hill top or some such thing that will line up with our boat to give a less formal check than the anchor bearings. Choose marks that will be visible in the dark.

If the exit from the anchorage is not an absolutely straightforward one, we make a note in the log of the safe course to take us out and clear of any nearby hazards. Then if we have to leave in a hurry, in the dark, we can do so without concern.

It is possible to drop an anchor when the boat is moving forward and if it is done properly the anchor will dig in satisfactorily but the technique can put your topsides at risk if things go wrong. *Didycoy* has an eighteen inch long bowsprit and the roller for the anchor chain is at its outboard end. This arrangement keeps the chain away from our topsides and gives us a third forestay as a bonus.

More often than not we would make our approach to an anchorage under sail. We would drop the jib well in advance and then, when we were getting in to anchoring depth, the main would be dropped to be held safely between the lazy jacks. As we coasted to a stop, head to wind, the anchor would be lowered and then sufficient chain veered as we backed away under the influence of the wind.

It gave us pleasure to anchor in this way and on those occasions when the engine had decided to have an off day, we were not thrown into a tizz because we had to anchor without it's assistance.

We have a few hand signals so that we can communicate with each other without shouting whilst we are laying or retrieving our anchor, so much more dignified!

## Tripping Lines

A tripping line is often recommended as an aid to recovering your anchor should it lodge under some obstruction on the sea bed. In use one end of a line is secured to the crown of the anchor .... the CQR has a ring provided for this purpose .... the other end is brought to the surface and buoyed. Now, if you experience trouble when you attempt to get your anchor aboard, the tripping line can be brought into action. First you must bring it aboard and then the anchor cable must be slackened. When this is done, hauling on the tripping line should pull your anchor out backwards from the obstruction.

In common with most passage makers we quickly gave up using a tripping line for two reasons. Firstly, when the anchor is being recovered, in the normal course of events, as you approach your anchor more and more tripping line is free to find its way to one's stern gear to foul the prop or rudder. Secondly, when lying at anchor with our usual generous scope laid out, the buoy would be some way ahead of the boat. In an anchorage off some idyllic third world island a passing fisherman might spot the buoy. With their limited possessions they are always on the alert to the chance of picking up something that could improve their lot .... I don't mean stealing .... a buoy apparently floating free would be fair game. When they find that there is something attached to it they haul it in. Their seamanship does not normally include a knowledge of the purpose of a tripping line. The chances are their anchor is a large stone with a hole in it .... no tripping line needed .... so there is no malice in their behaviour but there can be a large measure of inconvenience to say the least, for the skipper of the vessel as a result of their action.

Fortunately there is a very simple answer to the problem, it is the gadget illustrated in Figure 20 , a tripper. Haul short on the anchor cable so that it rises at a steep angle to the bow. Attach a length of line to the chain of the tripper and shackle the tripper around the anchor cable. Lower the tripper down the chain to the anchor so that it slides down the shank of the anchor. Slack off on the anchor chain and haul in on the tripper line, this should bring the anchor up stern first as it were, just as efficiently as a tripping line but with none of its disadvantages.

This simple gadget works so well and costs so little that there can be no excuse for not including one in your inventory. However, if you have failed to provide a tripper or perhaps lost it, it is possible to improvise. Use the heaviest ring spanner you possess in place of the mild steel bar you failed to provide and you are in business.

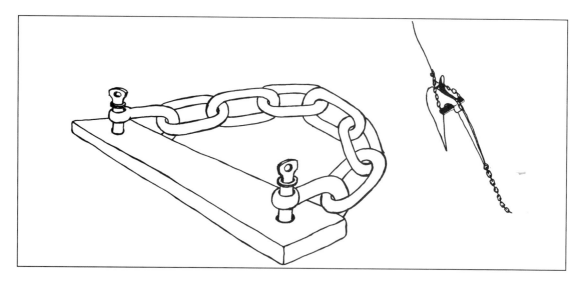

Fig 20. An anchor tripper

**Cable Marks**

Your anchor cable will need to be marked at intervals so that you will know how much you have let out. An easily remembered colour sequence is helpful and we have used red, white and blue at five fathom intervals. Whatever you do, record it in the log book as soon as you have marked the chain, it is so easy to forget precisely what you have done, or is it my extreme old age showing?

We first marked our chain with paint but found that it wore away far too quickly. It's a heavy task laying out 45 fathoms of three-eighths chain, measuring, painting and restowing it, so we resorted to red and blue plastic strips and short lengths of ¼in diameter line, they are still working for us. Failing a supply of suitable plastic strips, try using short lengths of line, one, two, three; one one, two two, three three, etc.

Our marks run :— a single red at 5 fathoms, white at 10 fathoms and blue at 15 fathoms. Double red at 20 fathoms, double white at 25 fathoms, double blue at 30 fathoms. At 35 fathoms we have red and white and at 40 fathoms we have red, white and blue. Two fathoms later we have a series of red tags to indicate that the end is nigh.

**Anchor Stowage**

The inboard end of the chain should be shackled to a length of strong line which, in turn, must be shackled to a strong point in the cable locker. This line needs to be strong enough to take the load should you inadvertently allow all the chain to run out. It must also be long enough to emerge from the navel pipe so that if you need to move from the anchorage in a great hurry you can buoy the cable and then cut the retaining line. If you do need to slip your anchor in this manner do use sufficient line to enable the chain to lie on the bottom but leave the buoy afloat!

When on passage most people unshackle the anchor from the chain and lash the anchor down very securely. It is not unknown for an anchor that is simply held by the brake on the winch to gradually pull some chain out against the brake and then swing back and fore doing great damage to the topsides before it is discovered.

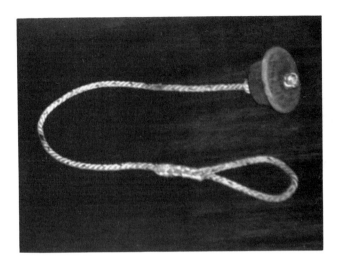

A navel pipe plug that will keep the chain locker dry.

We have a purpose-made wooden plug with a hole through it that takes a short length of line held in place by a stopper knot. The lower end of this line has an eye splice worked into it. With the anchor disconnected the eye splice is passed through the ring in the end of the chain and the wooden plug is passed through the eye splice to secure it to the chain. The wooden plug is then well greased and the anchor chain is allowed to drop down the navel pipe pulling the plug firmly into the pipe so preventing the entry of sea water should we take a green one on deck.

## Rowing Out an Anchor

If you need to lay out a second anchor it is usually done from your dinghy, and of course rope is used in this case. The anchor warp must be flaked down in the bottom of your dinghy so that it pays out as you row away from your boat. If you leave the rope on your yacht and attempt to tow it out to the point at which you wish to drop your anchor, you will find it a near impossible task.

Start by hanging the anchor over the stern of your dinghy and secure it with a slippery hitch so that a pull on the line will allow the anchor to fall free. Flake the rope down in the bottom of the dinghy and make the inboard end fast to a strong point aboard your yacht. As you row away the rope should pay out freely until you have it completely extended. Release the anchor with a pull on the slippery hitch. When you are back aboard your boat pull in on the anchor warp until you can feel that the anchor is safely dug in.

Whilst you are rowing the line out be sure to keep your feet clear of the rope as it pays out unless you wish to be 'marri-ed to the mer-mi-ade at the bottom of the deep blue sea'. Do not coil the rope down in the dinghy. That way it stands every chance of two or more coils snagging together and making an awful mess. Either flake it down in figures of eight or lay each successive bight at 90° to the last.

A deep notch in the centre of your dinghy's transom can be a help to recovering the rope laid for a second anchor. Without it your dinghy is apt to be pulled broadside-on creating difficulties and the possibility of capsize. If you don't wish to cut a notch in the transom of your dinghy perhaps you could bolt on a notched piece of timber when it is needed.

## Anchorages

At home we are accustomed to thinking of shelter as being an enclosed harbour or at least a river or estuary. Further afield one becomes a little less choosey. In common with the rest of the blue-water fraternity we have, on occasions, been grateful for the shelter afforded by a minute sandy cay or even a reef and the sooner you can come to terms with this aspect of cruising the easier will life be for you.

Of course bays like some of those found in the Marquesas, that have a comparatively small entrance offering all round protection, are the ideal, especially if they are as lovely as the Marquesas. Many tropical lagoons and atolls give a similar degree of shelter but they too are not always available.

There are three basic things we require of an anchorage. Good holding ground, shelter from the seas that build up with a strong wind and if possible shelter from the wind itself. Shelter from the wind is the least important of the three requirements but it is not to be scorned if it is on offer.

It is the build-up of the seas that accompany strong winds that makes for gross discomfort at an anchorage. Anchor too close to the seaward end of a piece of land that is to windward of you and the seas will roll round the corner and give you a hard time. With the boat rolling incessantly and probably pitching too, sleep becomes impossible and the slightest task requires a major physical effort. If you can tuck in somewhere behind a neck of land to a point where the seas cannot reach you, whether it be high or low, you will at least be protected from the movement of the sea and that's what really counts. If you are also protected from the wind consider it a bonus.

High land is not a guaranteed protection against strong winds. It is not at all unusual for the wind to increase its speed as it climbs over high ground in it's path and then blast down on the poor yachtie below. I can recall sitting at anchor at Lizard Island waiting for the wind to moderate. After six days we decided that we had waited long enough. We reefed right down, raised the anchor and departed. Once we were well clear of the island the wind dropped and we had to shake out all the reefs in order to sail. While we were doing this we gained a glimpse of the bay we had just left and the same old white-caps were rampaging across the water. I swear they were laughing at us!

Gloomy as it may sound, when you choose a spot in which to anchor ask yourself where you will fetch up if the wind turns and your anchor drags while you are ashore. If the answer is onto the only patch of rock on the beach, perhaps it would be prudent to choose a spot that would give you a softer landing.

# Chapter 9
# Anti-fouling

Haul-out facilities of some kind, ranging from travel-lifts to ramshackle railway and cradle outfits can be found scattered around the world. In the eastern Mediterranean many yards use a wooden sled and when the boat is settled on the sled the whole thing is hauled up a greased wooden track. As this involves the yard hands working in the water for part of the time it is easy to see why this method has not become popular in the North Sea.

The Indonesians have probably devised the simplest system yet, short of just beaching a boat. They ground their vessels over a huge log. One end of a stout tackle is attached to a strong point either side of the widest part of the hull at deck level. The other end is secured to the log and before the water can fall away the tackles are hauled up good and tight. Not a method to be recommended for fin and skeg vessels but any long-keel boat will sit upright in this way in complete safety between tides.

Some places have not yet come to terms with the modern racing machine and it's underwater shape and if you sail a vessel of this kind you will need to choose your yard with care. It makes sense to have a good photograph of your boat out of the water so that the yard hands can see what they are about to deal with. If it matters where the slings of a crane or travel-lift go, be sure that you know where these points are in relation to the upper deck.

This simple and effective way for any long-keeled boat to dry out safely. Note the heavy plank under the stern and the tackles made tight to hold her upright.

**Boat Legs**

Boat legs are not often seen these days but they still make sense, especially for a boat that is going to the more primitive parts of the world. True, a reasonable rise and fall of tide is needed to expose a boat between tides and much of the tropical world lacks tides of a suitable size. Nevertheless there are enough places where legs are usable and you only need to save the cost of one haul-out to start making a profit and what is more you can do the work on a nice quiet beach of your choice and not in some noisy, dirty boatyard.

Traditionally boat legs have been made of timber and to gain enough strength they have needed to be heavy and therefore cumbersome to stow and to use. Typically timber with a cross section of 6in x 4in has been used for vessels between 36ft and 45ft length overall and $3^{1}/_{2}$ x $2^{3}/_{4}$in for boats up to 25ft. Steel tubing of the kind used for scaffold poles can be used in place of timber and this will ease both the handling and stowing of a pair of legs.

The length of the legs will be determined by the distance from the bolt hole in the side of your boat and the ground when the boat is dried out. The legs must be slightly shorter than this distance to allow for the fact that the weight of the boat will cause the keel to sink into the beach a little. Opinions vary on just how much shorter they should be, which suggests that absolute conformity to a particular figure is not vital. The limits I have seen set, range between 1in and 3in; we made our legs 2in shorter and they seem to work well enough.

To prevent wooden legs splitting when subjected to load there must be at least two feet of timber above the bolt holes. Steel legs will not need to extend beyond the bolt holes by anything like this much. We have welded to the top of each leg a semicircle of $^{1}/_{4}$in diameter mild steel rod, so that a line can be secured to it to help support the weight of the leg whilst it is being rigged.

*Didycoy*'s legs are made of 2in inside diameter galvanised steel tube with a wall thickness of 5mm. The fore and aft position for the legs needs to be at the widest point of the boat and a stiffening pad must be fitted to the interior of the hull to spread the load. In a wooden boat a hardwood pad would be the obvious answer. In a glass fibre boat, a large timber pad or a galvanised steel plate could be glassed in place. The plate would need to be drilled to take the bolt before it was galvanised. The bolt hole in a timber boat should to be about three planks down from the deck level. When the legs are not in use a short mushroom-headed galvanised bolt is used to plug the hole.

Our legs are held to the side of the boat by eighteen inch long steel tubes that are just large enough to slide over the legs. A seven-eighth inch diameter steel bolt, with it's head cut off, is welded to the mid point along each short tube. It was considered prudent to reinforce this point of attachment and so when the bolt was welded in place a 4in length of the same tube was split longitudinally into three pieces, we needed two of them. With a large hole drilled into the centre, one of these plates was dropped over each bolt and then welded to both the sleeve and the bolt.

In use the leg must stand vertically and to achieve this a shaped wooden pad is fitted between the hull and the leg in the region of the bolt hole. One side of this chock must be shaped to provide a trough that will take the sleeve and hold it vertical and the other side should be shaped to fit snugly against the side of the boat in the vicinity of the retaining bolt hole. If this chock is padded your topsides will be protected from damage. The chock must be drilled to take the bolt on the short tube and allow it to pass through the stiffening pad to the inside of the boat where a washer and a nut are used to secure it to the hull.

*Didycoy's* Legs

With ease of stowage in mind, separate feet were made for the legs. A 6in square of ¼in steel plate with a 4in length of the larger tube welded to its centre forms each foot. If you don't consider it an impediment to easy stowage, the plate could be welded directly to the leg. The foot plate has been drilled to accept a bolt in each corner. This allows me to secure a 12in x 12in pad of half inch thick ply to each foot. It was thought that a steel plate of this size might make the foot too heavy for easy handling. Holes in the foot tube and the bottom of the leg make it possible to bolt them together with five-eighth inch bolts.

Not all boats are suited to the use of legs. The legs are intended to keep your boat upright whilst she sits on her keel. If your vessel tips onto her nose when grounded or has a knife-like fin keel, then legs are not for you and it serves you right.

In use the prime requirement is a good firm level area; soft mud and the like will only cause problems. Carry out a survey at low water and when you have found a suitable spot either note or erect marks that will lead you in at high water. Check that the next tide will be big enough to lift you off again!

When the time comes to rig your legs, fore and aft guys must lead from the foot of each leg to bow and stern and these must be tight. When you motor ashore go gently. That way you will be less likely to knock the legs out of plumb. Should that happen it is usually possible to reset them by hauling on the appropriate guy. Put out a stern anchor at some time or an onshore wind could cause trouble when you try to get off.

I can well recall the trepidation with which we approached our first drying-out attempt. There was no one to lend us moral support, we felt very lonely indeed. In the event we crept ashore and the whole thing virtually worked itself. It really was as easy as that. Problems can arise of course, but if you have chosen your ground well and you are only going to be out for one tide you should have no trouble. If you intend to stay out for more than one tide and expect to lift at each high water it is prudent to shift your position sideways at least two boat widths.

On one not easily forgotten occasion we secured *Didycoy* with a bow line and a stern anchor assuming that we would settle back into our original position. We didn't, we moved sideways a little and when we settled we developed an alarming list. We hadn't moved just anywhere, we had moved exactly half a boat width and one leg sat precisely in the six inch deep rut made by the keel on the previous low water. There was a panic stricken rush for a couple of logs that were just the right length to stand on end and jamb under the rubbing strake. A halyard was extended and taken out from the masthead to a nearby coconut tree and there we stayed, rather nervously.

When lift-off time came with the next tide we stayed put. This was, to say the least, rather disturbing as with the peculiar tides we were experiencing we had only the next tide to lift us off or stay in that unhappy position for three more weeks. When the tide had receded I examined the offending leg. It had dug in to a depth of about a foot and sand had washed in and compacted to the level of the surrounding ground. I dug the sand out and continued digging as the tide rose until I was forced to wear goggles and snorkel and finally to give up.

If we had failed to lift I had ideas of heeling the boat to reduce the draught by hauling on the masthead rope. There was no need. Murphy knows when he is on to a winning streak. He allowed us to lift off the beach and as I backed off into deeper water I wrapped the anchor rope firmly around the prop.

## Anti-fouling Paints

Before we left the U.K. I had read that it paid to use the same brand of bottom paint as the local fishermen. The argument was that fishermen were no fools and would know which anti-fouling paint gave the best value for money in their area. When we reached the tropics we soon learned that most fish in those areas are caught in the vicinity of rocks and reefs. The result was that fishermen liked to inspect their hulls for damage at frequent intervals and so bought the cheapest anti-fouling paint they could find. As the man said, they are not fools.

Hauling-out is not a cheap pastime so clearly it pays to use an effective anti-fouling to increase the time between slippings. The best one can say of the anti-fouling paints that are offered for sale is that some are even less effective than others. I find it difficult to get excited about the performance of any of them. Merchant ships expose considerable areas of their hulls to the air and sun for quite some time between discharging and loading cargoes. Their hulls never appear to be fouled with marine growth and they manage to go for two years or more between cleaning off and repainting.

When we hauled out in Saudi Arabia, the yard painted our underwater section with big ship anti-fouling. Thoughts of that paint job lasting us for two or three years were soon shattered. Six months later we were quite badly fouled up. I can only think that our slow speed, compared to that of the average ship, was to blame for the poor performance of the anti-fouling paint that had been used. This thought was reinforced by the difficulty I had in persuading the yard hands to paint our propeller. It seems that the only treatment given to a ship's propeller is to polish it. A yacht's prop is idle for a very large part of it's life but a ship's screw is moving at a fast pace for most of it's time.

Soft anti-fouling paints tend to rub off far too easily. A mooring line in the water or some similar obstruction can rub off a lot of soft bottom paint in a short space of time. If you dive to scrub off growth from this kind of paint the chances are that you will take a large amount of paint off too.

In the past many soft anti-fouling paints had to be immersed in water within a specified period of time after the completion of painting, usually twenty four hours. If you then found it necessary to haul-out for a repair of some kind before the hull needed repainting the additional exposure to ultra-violet light destroyed what anti-fouling properties the paint had left, probably in a matter of hours. Many soft anti-fouling paints have been improved on this point and they can now be exposed to sunlight for up to a month or so. If you wish to use soft anti-fouling, check the label to be sure that you have the right one. My other objection to soft anti-fouling paint remains valid.

There are a number of materials that can be applied to a wooden hull which, it is claimed, will keep worms at bay. How effective they are I don't know but the fact that some of them dictate that a soft anti-fouling must be used puts them out of court so far as I am concerned. The West™ resin saturation process is a different proposition and merits serious consideration, but it is only applicable to new wooden construction.

When we were in Cyprus, many yachties were enthusiastic about the effect of 2.5 gms of Tetracycline in each litre of their anti-fouling paint. They claimed that it kept a boat free of all but a slight slime for two or three years. The examples I saw certainly confirmed these claims. This is a practice that the medical profession is not the least bit happy about. In this country one would need a prescription to be able to buy any form of antibiotic. In Cyprus, and in many other countries too, drugs of this kind are freely on sale over the counter and I was somewhat amused to note that, typically yachtie, they had realised that Tetracycline from a veterinary surgery was far cheaper than that bought from a pharmacy.

Another promising development that, so far as I know, does not languish under anyone's disapproval is the use of powdered copper in resin. Powdered copper is used by sculptors to give glass fibre the appearance of copper. When the work is laid up, powdered copper is incorporated in the gel coat. This method has been used in the lay-up of some boat hulls and at a later stage the hull is scoured with wire wool to expose the copper. I have no experience with the method but a friend whose judgement I trust has applied a coat of copper-impregnated resin paint to the hull of an existing glass fibre boat and scoured the surface when the resin was hard. The hull has remained free of growth for two seasons and there is no reason to believe that it will not continue to be effective.

The materials are not cheap but then neither are anti-fouling paints. If you wish to try the copper/resin paint, you will need to buy the following materials:—

Sufficient epoxy resin and catalyst to cover the underwater surface of your boat.

Enough copper powder to give you six to seven times the weight of the epoxy resin.

Styrene to act as a thinner to the epoxy resin.

The styrene needs to be added slowly to the epoxy until the resin reaches a consistency that will allow you to paint it on. The copper powder must be mixed with the epoxy. If you aim to have a glass fibre boat shot blasted and painted with a new gel coat, it could be your chance to incorporate some copper powder in the final coat.

Check with the supplier on the working time you can expect with the mixture you intend using. The higher the ambient temperature the shorter will be the period in which you have to work before the mix becomes too hard to spread. 60°F or 15°C is usually considered to be the lowest temperature in which resin should be used; lower than this and the resin will be reluctant to 'go off'. When the resin has hardened you will need to rub the surface down with wire wool to expose the copper.

Look - no hands ! The support is actually on the far side of this schooner.

# Chapter 10
# Heavy Weather

From conversations we have had with potential blue-water sailors it would seem that bad weather and lack of weather forecasts are matters for some concern. The availability and reliability of weather forecasts around our temperate home waters have led many yachtsmen to rely heavily on them. This makes good sense in the limited and congested waters around well populated shores. I would not want anything I write to change that attitude in those waters. It becomes a very different matter when one ventures further afield.

With the exception of a few places, the waters cease to be congested, indeed on the ocean ships seem to be a rarity. I accept that there are many thousands of ships at sea at any one time but the oceans of the world are so vast that the ships are literally a drop in the ocean. Once we were away from the land and sailing the trade wind route to the West Indies we spotted two ships in twenty one days. If you don't feel humble and insignificant before you sail an ocean or two, you will before you finish, that's for sure.

The sheer size of the oceans usually means that you can stow all sail and safely heave-to or run before the wind for a great many miles in any direction you choose if you encounter really bad weather. Hardly something you could do in home waters. One soon begins to realise that the immensity and the emptiness of the oceans is a safety factor in itself. There is an ocean of truth in the expression, so often heard when in the company of fellow passage makers, that 'if I can see it, it is too close'.

For most of your time you will be sailing in areas where weather forecasts just don't exist but, so long as you choose the right season, the weather is likely to be consistently good. On those few occasions when you suffer really bad weather, as long as you ride the punches and do not try to fight nature and you have a boat that is designed to go to sea, the suffering will not be unbearable.

Naturally you cannot expect to swan around the oceans of the world for several years without meeting some bad weather, but it seldom lasts for more than three or four days and later you will look back on it as just one more experience that added spice to your voyage. It would be great pity to let thoughts of meeting heavy weather spoil your anticipation of the tremendous pleasures that are in store for you. With a seaworthy boat, as much understanding of heavy weather seamanship as you can gain from reading, allied to your past experience and all the sea room you could want, you have little to fear.

Before we left Britain in 1983 we found ourselves stormbound in the attractive little haven of Newton Ferrers, quite close to Plymouth. It is in the mouth of the River Yealm and sits in a deep cleft in the rocky hillside facing west. The wind was from the south west and blew straight in virtually unchecked, but we were spared the effect of the seas rolling in from the Western Approaches. The wind speeds reached eighty knots for quite long periods. Hurricane force winds start at sixty four knots so I suppose we can claim to have experienced Hurricane strength winds before we left the UK, but we were in harbour and securely moored to a large concrete pontoon in midstream.

The surface of the water was lifted as a dense mist to a height of about forty feet and driven before the wind up the estuary. This mist arrived with such force that it was not possible to face it. Several multihulls of about twenty five feet or so were overturned at their moorings and other boats had their sails ripped from their gaskets and shredded in minutes. A lightweight glass fibre boat moored onto the pontoon to which we were secured was leaping about like a demented thing and it repeatedly attempted to mount the pontoon. At the height of the storm, which lasted for about five days, conditions were so bad for the couple who were aboard that we took them into *Didycoy* for forty eight hours.

This limited experience most certainly sharpened our respect for the danger that lies within these massively powerful weather systems; especially so when you realise that one hundred and forty knots is often reached and sometimes exceeded, almost twice the wind speed we experienced.

## Tropical Revolving Storms

Tropical Revolving Storms (TRS) are given a variety of names depending on the area in which they appear; Hurricanes in the West Indies, Hurricanes or Cordonazo on the eastern side of the North Pacific and Typhoons or Baguios on the western side. In the South West Pacific and the Indian Ocean the name is Cyclone. Whatever their name they amount to the same thing, a deep low pressure area with winds in excess of sixty four knots.

Tropical Revolving Storms usually occur in the latitudes between 10° and 20° North or 10° and 20° South. It is not unknown for them to move out of these latitudes but as a general rule it holds good. The areas afflicted by Tropical Revolving Storms are fairly closely defined as is the time of year in which they occur.

To be able to develop, these storms require a well heated area of sea and a cyclonic disturbance. An island could provide just such conditions. Land heats more readily than the sea and so the rising air currents will be accelerated by this unequal heating. The constantly rising air must be replaced and so horizontal air currents come in above the sea and we experience them as wind. The faster the air rises the stronger are the winds to which we are subjected. The rotation of the earth imparts a spin to the rising column of air. In the southern hemisphere the rotation is clockwise and in the northern hemisphere it is anti-clockwise.

An average TRS is about 500 miles across but they come bigger, 750 to 800 miles in diameter, and smaller, perhaps as small as 50 miles across. Because a TRS is dependant on a continuing supply of warm moist air to keep it in being it will die out if it travels far enough across land, but not before it has wreaked havoc on the coastal belt. It is not unknown for a storm to regain the sea and when that happens it usually intensifies again.

To return to our rising column of air. As it rises so the ambient temperature falls and the pressure drops at a steady rate which allows the air to expand. An expanding gas will cool and these cooling processes eventually lower the temperature of the air to its dew point. When this point is reached the air can no longer retain its moisture as invisible water vapour and gives it up as cloud. When the water droplets within a cloud unite to make drops heavy enough they fall against the rising air currents. If the drops of water freeze on the way down and remain frozen they fall as snow or hail but in the tropics it falls usually as exceedingly heavy rain, occasionally as hail.

The path followed by the perfect TRS will take the form of a parabola.(Fig 21) In both northern and southern hemispheres it will initially move westwards. North of the equator a cyclone will move a little north of west, about WNW until it recurves northwards and then it will move to the northeast. In the southern hemisphere it will start with a west-southwesterly path, later it will curve to the south and then to the southeast. The initial speed of a Tropical Revolving Storm will probably be about ten knots, it will slow as it

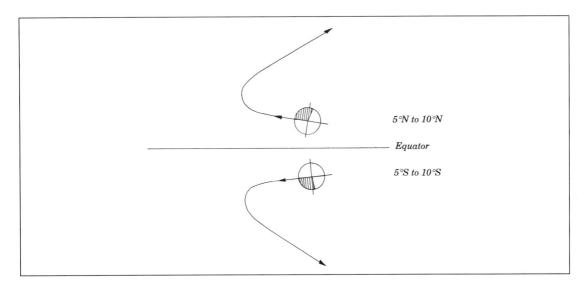

Fig 21. The paths of Tropical Revolving Storms in the northern and southern hemispheres.

recurves and then the storm will pick up speed again reaching perhaps as much as twenty knots.

You will have noted the probabilities and the perhapses and also the reference to the perfect TRS and doubtless you will have drawn your own conclusions and will not be at all surprised if your very own TRS deviates from the normal pattern. Our real interest is in how to avoid all Tropical Revolving Storms or failing that, how to handle the problem. Avoidance is simple. Stay well away from TRS areas in the appropriate seasons.

Many passage makers crossing the South Pacific from East to West linger in French Polynesia for the cyclone season, and what could be better? The Marquesas and the Tauomotus are beautiful cruising grounds. Tahiti is usually considered to be the furthermost west it is prudent to venture between December and April. If you must press on then the early and late parts of the season are usually the best bets. New Zealand is another justly popular area to explore whilst cringing from the nastiness further north. As December to April is the summertime in the southern hemisphere you will be enjoying the best of the weather in a cooler climate. Both Tahiti and New Zealand have extensive facilities to enable you to carry out a major refit

The next best plan, if you must keep moving, is to explore somewhere like the Queensland Coast of Australia making your way north along the Great Barrier Reef with one ear on the radio weather reports and one eye on the next cyclone hole, but it is a nervous way of sailing.

Your radio is one source of information and your own ability to observe and interpret is the other. Barometric pressure fluctuates very little in tropical areas; the pilot book will tell you what to expect the barometer to read and the normal diurnal variation to expect in your area. Anything outside this range will suggest trouble. In many tropical areas a fall of more than three millibars would be considered significant. With an approaching TRS the pressure is likely to be abnormally high at first and then it will drop ..... and drop.

Bands of nimbo cumulus with some thunderstorms will appear in the early stages until the sky is covered with a solid deck of cloud. Moving fast under this solid deck of cloud will be vast areas of ragged torn off clouds. The total clouds will indicate the position of the storm. The direction in which the lower ragged clouds are moving will indicate the path of the storm. Again I must admit that we found it difficult to decide on the direction of cloud movement at sea. Sunrise or sunset may appear to be a sickly greenish yellow.

When changes such as those detailed above are observed aboard a yacht it is probably too late to avoid the rising storm. Merchant ships with three or four times our speed may have the ability to escape the worst of a TRS by steaming away from it, but our speeds are such that we have little hope of out distancing one. This is not to say that nothing can be done. The greater your understanding of the behaviour of your boat, the weather and the sea the better are your chances of survival.

The direction of the wind you are experiencing will indicate the direction of the centre of the storm relative to your position. In the northern hemisphere when the barometer starts to fall the centre of the depression will be about 130° to your right when you are facing into the wind. When the barometer has fallen another ten millibars the centre will be about 110° to your right and when it falls a further ten millibars the storm centre will be about 90° to your right.

Because the wind rotates in opposite directions in the two hemispheres you must change 'right' to 'left' in the paragraph above when referring to the southern hemisphere.

A swell that runs counter to the normal pattern could be your first warning but we have found it very difficult to detect. Until recent years most information on Tropical Revolving Storms was collated from reports submitted by Masters of merchant vessels. I suspect that from the lofty perch of a ship's bridge a swell that runs counter to the normal pattern could be detected. From the cockpit of a yacht with an eyelevel of some six feet it is not so easily seen.

WWV, the American radio station that broadcasts continuous radio time signals, also gives Tropical Revolving Storm warnings and their progress at twelve minutes to every hour. The same station gives a nil report which can be quite a comfort if you are tip-toeing across an ocean at the wrong time of the year. A broadcast TRS warning will usually give the position of the centre of the storm, its diameter and its expected path and speed.

If you learn from your radio that a storm is building in your vicinity, plot the centre of the TRS on your chart and on this point draw a circle of the reported diameter. Plot the predicted path of the storm through the centre of the circle and extend this line backwards a little way. From this line draw two lines tangential to the circle to make an angle of 40° either side of the line of advance. This arc of 80° will cover the area of possible deviation from the predicted path, Figure 22A. With this information plotted you will be able to see where the hazards lie in relation to your position and equally important where you lie in relation to the storm. Do not underestimate the importance of plotting this information and keeping it up to date. There will come a time when all you can do is react to whatever conditions are confronting you, but that will be later.

Now you can decide which way to move to improve your chances of survival. The general rule in the northern hemisphere is that the best course to take you away from the storm will be that which puts the wind on your starboard quarter. In the southern hemisphere you will need to put the wind on the port quarter. Like all general rules this one must be tempered with other information and commonsense.

The two leading quadrants of the storm are called 'the dangerous quadrant' and 'the navigable quadrant', Figures 23 A & B. The 'dangerous quadrant' is so called because the wind's speed is increased by the speed of the advance of the storm system and the 'navigable quadrant' winds are reduced by the speed of the advance of the TRS.

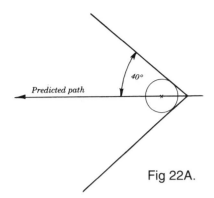

Fig 22A.

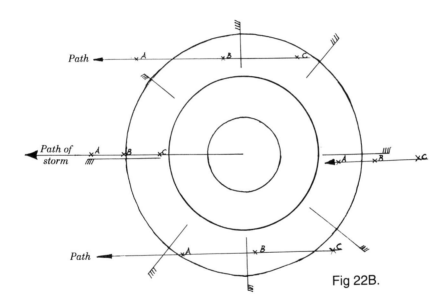

Fig 22B.

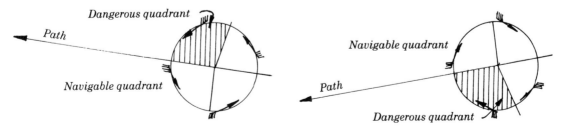

Fig 23A.  *TRS in northern hemisphere*

Fig 23B.  *TRS in southern hemisphere*

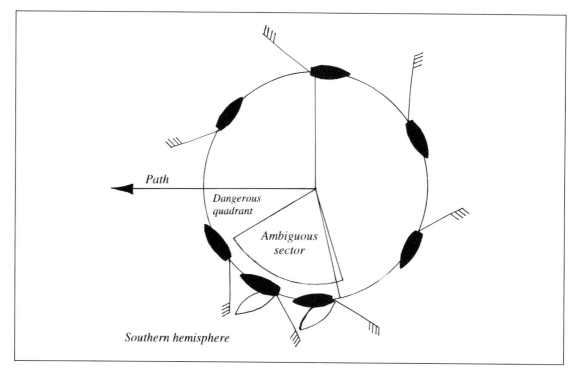

Fig 24

If you examine Figure 24 with the advice to put the wind on the appropriate quarter in mind, you will see that such a move may not be ideal if you are in the dangerous quadrant. There is a sector of this quadrant of about 65° within which the standard advice would drive you deeper into the TRS. I am not aware that this sector has a name, so for ease of discussion I will call it the ambiguous sector.

Ships that find themselves in the ambiguous sector, or in its path, are advised to steam with the wind on the port bow in the southern hemisphere and with it on the starboard bow in the northern hemisphere. This will present them with the shortest route away from the centre of the storm. The same tactic would serve for us but we do not have power enough to motor against winds of the strength we are discussing.

I suppose we had about three days warning of the approach of our own cyclone and for much of those three days the winds were very strong but they were manageable. This does suggest that providing you take early action there is every hope that you could move away from rather than towards the worst of the storm.

This last sentence sums up what you must attempt to do in the early stages. There is no way that you can hope to escape several days of very strong winds but a nine is better than a force twelve and a seven would be even better. The seas will not have built up enough to stop you at this early stage and the winds may well not yet be so strong that it is impossible to attempt to sail out of trouble.

If you judge that the ambiguous sector is headed your way, you will still need to get the wind on the appropriate beam but now the sheets must be hardened in as far as the wind and sea will allow to take you away from the path of the dangerous quadrant.

Area G on the world map on page 90 indicates an area where the winds of some cyclones can blow straight across the isobars towards the centre of the cyclone (Fig. 22B, p.79). This would make it impossible to estimate the direction of the centre of the storm by the above means. In this special area the wind direction and the behaviour of the barometer can still be very informative. Should you find that the wind direction does not change with the passage of time and the barometer continues to fall then you have the misfortune to be directly in the path of the centre of the storm. On the other hand if the barometer continues to rise with an unchanging wind you are behind the centre of the storm which is heading away from you.

A wind that changes its direction in an anti-clockwise manner, for example from 300° to 270°, is said to back. One that changes in a clockwise direction is said to veer. If your observations tell you that the wind is backing you must be in the left hand semicircle. A vessel in the right hand semicircle will experience a wind that veers as the cyclone passes his position.

To sum up.

1. Plot the radio information about the position etc of the storm.

2. Check the wind direction against the barometric fall to confirm the position of the centre of the TRS as given by the radio. If your observation differs from the radio forecast, trust what you can see.

3. Do your utmost to move away from the storm's path into an area of less extreme conditions.

All this indicates a need for a very real understanding of the signs, symptoms and behaviour of tropical revolving storms. Without this understanding you will not be able to plan your action early enough to ensure your safety. Our storm experience has been confined to the tropics and the subtropics. If it is your intention to go down to the Southern Ocean then you must study the reports and comments of people who have been there.

## Heavy Weather Options

There are several choices open to a boat that is experiencing extreme weather conditions. Which tactic is used will depend on a number of factors. If circumstances permit it clearly makes sense to keep moving away from the centre and/or the path of the storm for as long as this is tenable by sailing with the wind on the appropriate quarter.

It may be that in the light of your particular circumstances you will decide that you do not wish to run but would prefer to limit your downwind progress as much as possible.

There are two terms that we must use and it may be helpful if I define what I mean by them.

## Heaving-to

This simply means to bring a vessel to a halt and is usually achieved in a sloop by backing the jib and lashing the helm to keep the vessel's head to windward. In this condition most boats will lie quietly unattended, sailing a little to windward and drifting slowly to leeward .... one movement partially cancelling the other. With a boat jogging along quietly in this way the cook can work more easily or a problem can be sorted out more readily than if a boat was racing along.

From this description you will see that it is not a severe weather gambit, although a suitable boat can use a modified version, as we discovered. Bearing in mind that what is manageable weather in one place can be heavy weather in other circumstances, heaving-to under sail can sometimes be used to advantage in difficult conditions. It may be that you are beginning to suffer a confused sea with the promise of more turbulent conditions ahead, but at the turn of the tide you expect the sea-state to improve. With a wind of perhaps no more than a force six or seven a storm jib and a deep reefed main might well let you lie in relative comfort until the tide turns to your favour.

Heaving-to under sail in anything stronger than a force seven or eight could well prove to be too much for your gear. In the open ocean one would expect to run happily before a six or seven.

### Lying Ahull and Lying Atry

Two names for the same operation. Lying ahull is the term most often used.

Lying ahull requires that all sail should be stowed and the yacht allowed to find its own position in relation to the wind and sea. The most usual position adopted is beam-on to the advancing seas.

I dislike this tactic for a number of reasons, not least because it is tantamount to opting out and hoping that your boat will take care of the situation. It is a negative approach and I feel most strongly that all that you do when confronted with a serious problem should be positive, if only for the sake of everyone's morale. The other drawbacks to lying ahull will be discussed shortly. Rather than abandon control of the boat I would prefer to attempt to heave-to with all sail stowed. We found that *Didycoy* would heave-to in this way with the helm lashed to keep her as close to the wind as possible. On one occasion in the South West Pacific we sat like this for four fearsome days. *Didycoy* lay at about 50° to the wind and rode every sea that came her way, taking no great weight of water on her decks at any time. Whilst the movement was considerable, and at times violent, her behaviour was always reassuring.

For large parts of those four days the sea was carpeted with a dense pattern of white water. At times the breaking seas were, in part, subdued by the immense weight of the rain that fell from the skies only to reform once the rain had stopped. It was at times like this that we were grateful that we had chosen a boat with a long and heavy keel. There is plenty of evidence to show that a lighter vessel with a fin keel would not behave in this reassuring manner. No matter how you arranged the helm, a fin-keeled vessel would probably lie beam to the seas as if it were lying ahull, exposing her crew to considerable movement which will sap both efficiency and morale. It would also expose the boat to two very real dangers.

In a severe storm the waves build and become steeper with time. At some point a boat lying beam on to the seas will be in great danger of falling violently from the crest of an advancing wave into the next trough down wind, Figure 25. Falling to leeward like this can cause considerable structural damage to a boat. As if this were not enough, the same set of circumstances can readily lead to a capsize. With a boat lying beam on to the seas she will be heeling quite badly especially if she lacks a heavy keel, at times to the point where the lee scuppers are submerged. While she is in this position the next wave to arrive is likely to strike the hull and attempt to drive her to leeward. The lift of the second wave and the resistance of the submerged scuppers could well be enough to cause the boat to rotate and capsize, Figure 26. If it becomes obvious that you are approaching either of these dangers you must change your tactics and start running before the wind.

When we hove to for those four days we were on passage between Tonga and Australia, about 250 miles east of Brisbane. The wind was such that with it on our port quarter and

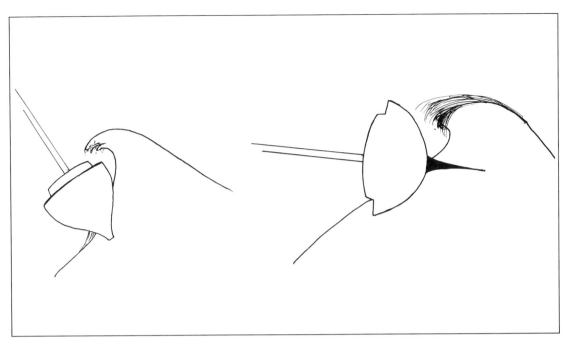

Fig 25                                                Fig 26

under bare poles we would have been heading back towards Tonga at a high rate of knots, something we were most reluctant to do. The wind direction had convinced us that we were on the northern side of the storm and that our course of action would allow the centre of the storm to pass to our south without us sacrificing too much of the progress that we had made in the previous two weeks. With all sail stowed we put the wind on the port bow and lashed the wheel right up to windward. This held us about 50° off the wind, in short we were hove-to under bare poles.

Although lying at 50° to the wind is not quite the same as sailing at 50° to the wind, *Didycoy* rode the seas in a most reassuring manner. She undoubtedly sailed slowly to windward which countered some of the leeway the violent conditions were creating. When the skies cleared enough to allow me to use the sextant to fix our position we found that our drift proved to be at about 180° to the wind and in the four days that we lay hove-to we made 100 miles to the east. Not only did we save perhaps 300 miles by heaving-to rather than running, but we also sat comfortably, comfort being a relative term! We were able to sleep instead of taking our turn at the wheel in an effort to ensure that *Didycoy* was able to safely take every sea that approached us from astern. At no time did we take green water on the deck, nor were we ever knocked down.

For much of the time the air was dense with wind-driven spray. From time to time the rain was so heavy that it subdued the breaking seas. By this time we were well accustomed to the weight of tropical rain but this exceeded anything we had met before. Not only were the breakers subdued but the very seas themselves seemed to be reduced by the sheer weight of water that fell on them. The noise down below was incredible.

Once heaving-to or lying-to becomes untenable then a boat must be allowed to run before the wind. If bare poles do not give enough speed to maintain directional control, especially in the troughs when the wind may not reach much of the boat and its mast, then you will need to raise a scrap of sail.

If I were ever again to venture into waters where I might meet winds of force ten or more I would take a very small trysail that could be raised on the main halyard and a similarly small storm jib. I think the design should be tall and narrow with a very high cut clew. We could most certainly have made good use of a pair of such sails in our battle up the northern half of the Red Sea at the wrong time of the year. The standard storm trysail we had was too big for those conditions and we spent much of that passage beating under our storm jib alone.

## Running Before a Storm

The need to run before the wind will arise sooner with some designs of boats than with others and you must be the judge of when that moment arrives. If you are forced to run get the wind on the appropriate quarter, this will do two things for you. Not only will it take you away from the path of the storm but it will present the cheek of the bow of your boat to the water ahead giving more support than a truly head-on approach will give and so reduce the chances of pitchpoling.

When running, the dangers of capsize are greatest when the crest of the wave is midships. It is exacerbated when the speed of the boat is such that she lingers in that position. A large heeling moment applied by wind or sea, or both, whilst a vessel is in this position will induce broaching and/or capsize. Clearly if you find yourself in this situation it would make sense to change the speed at which you are running by whatever means is available to you.

Wave effect is greatest at the surface and diminishes fairly rapidly with depth.(Fig 27) This means that a flat bottomed, shallow draught vessel will suffer more than a deep draught vessel. A shallow hull with a deep keel does not does not qualify as a deep draught vessel in this context, the hull must penetrate to deeper water for a real advantage to be gained.

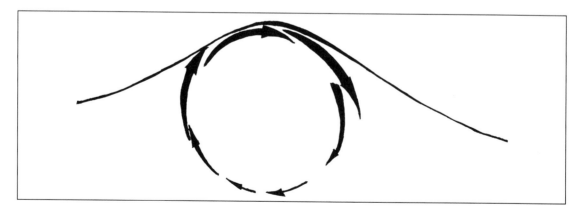

Fig 27. Wave effect diminishes with depth.

**Streaming Warps**

Warps have been towed astern to reduce speed and to help keep the boat on course before the wind. It is now a well understood tactic and seems to be a valuable and worthwhile move now that the problems have been ironed out. In the early days it was thought that a heavy line suitably weighted with car tyres, heavy anchors and the like was needed. When it was found that a following sea would pick up this gear and hurl it into the cockpit the idea lost some of its popularity. Fortunately for us some people persevered and modified their approach until it was realised that weight was not needed. Buoyant line was found to work well if it was used in sufficient length.

One hundred fathoms of fifteen millimetre diameter Polypropylene is recommended as the most suitable warp to tow. It should be towed in a bight with an end secured to each quarter. Be sure to slip a stout piece of plastic tubing on to each end of the line to protect it against chafe where it comes into contact with the boat. Make sure that the plastic tube is secured to the line or the boat or it will move and allow the line to chafe through in no time at all.

C.A.Marchaj has backed up earlier practical work with some extensive theoretical study. He recommends the use of buoyant warps in the earlier part of the storm if a boat needs to run. As the storm develops so the seas shorten ....get closer together....and there will come a time when you will need to dispense with the towed warp.

**Sea Anchors**

Sea anchors have been fashionable but their popularity has declined considerably in recent years. They are designed to hold a boat's head to the oncoming seas. Modern fin-keel boats seem to hang anywhere from 60° to 180° from the wind and sea which is not the idea at all. To my mind a sea anchor pins a boat down when it should be free to ride with the violence the sea offers. There is nothing more foolhardy than standing still so that you can take it on the chin. (Figure 28.)

As if this was not bad enough the very design of the sea anchor invites problems. The anchor consists of a wide mouthed cone of canvas that is held open by a steel hoop to which is attached a bridle. The tip of the cone is open and supported by a grommet of rope to which is attached a tripping line. Also secured to the bridle is a single sheave block through which is reeved an endless whip. The endless whip is there to allow you to haul out a canvas bag filled with fish oil to calm the raging seas. When the bag of fish oil is empty you haul it back inboard and refill it. Naturally it must be fish oil .... where you buy that in this day and age I can't imagine.

I don't know about anyone else but I have given up towing two fishing lines in good weather because they always become inextricably tangled within an hour of streaming them. What state a warp, an endless whip, a tripping line and a bag of fish oil would be in after an hour of bad weather I shudder to think. The four lines would be so intertwined that they surely must have been the inspiration for Multiplait. As for the state of one's foredeck with all that fish oil about.... .

Twenty four foot diameter parachutes are being sold as sea anchors and it is said that they are very efficient. Be that as it may I would far rather be free to ride the seas than be helplessly tethered by the nose.

And finally I would say, 'Hang in there, nothing lasts for ever and bad weather is no exception'.

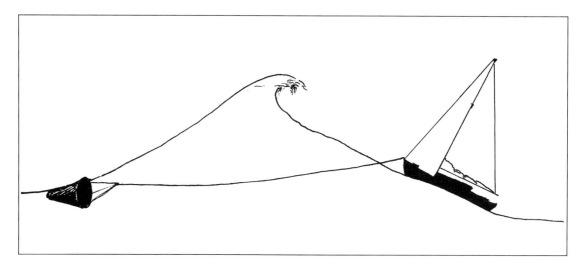

Fig 28. The perils of the sea anchor.

## Storm Tactics in Harbour

What if you are in harbour when warnings of an approaching TRS are broadcast? The oft stated advice that you would be safer at sea than in a crowded harbour will take a good deal of courage to carry out, but if circumstances are right it could be excellent advice. In addition to courage it will require a good sea boat and plenty of sea room. If you are to leave the sooner you leave dubious shelter the better, there may well come a time in the course of a storm when conditions are so bad that you no longer have the option.

In some circumstances it may be foolhardy to attempt to put to sea. If the storm winds are blowing onshore, clearly a crowded harbour could just possibly be a better gamble than the lee shore awaiting you outside the harbour mouth. If, as on the North Queensland Coast for example, your route to the open sea is blocked by the Great Barrier Reef or some similar obstruction there would be no point in attempting to head seawards.

Harbours are often situated at the seaward end of a river and despite the undoubted beauty of many rivers they are nothing but a drain for the surrounding land. The rain accompanying a Tropical Revolving Storm is usually so heavy and so widespread as to cause serious flooding. The TRS that struck Queensland in 1990 inundated an area equivalent to that of the British Isles. In 1992 a TRS in the same area deposited twenty seven inches of rain in twenty four hours!

The extremely low barometric pressure that accompanies these storms allows the sea and rivers to rise substantially simply because the weight of air resting on the water's surface is so much less than normal. When this is added to the vast volumes of flood water that will find its way down river it is easy to see that the depth of the river will rise considerably and the current will accelerate to a high rate of knots. Trees will be uprooted, telegraph poles and a wide variety of timber structures will be blown down. Much of this and more will make its way down stream at great speed.

These three factors will combine to endanger the security of nearby anchored vessels and those that drag their anchors will add to the hazards offered to those who are still anchored. In these circumstances there is a distinct possibility of finding your boat at rest on the Town Hall steps when things have calmed down.

Naturally this is not all that could happen. Winds of these strengths are so destructive that flying debris is a very real danger. Many buildings in the tropics are roofed with corrugated iron sheeting and it takes only a small effort of imagination to envisage the lethal effect of a squadron of corrugated iron sheets flying down the main drag at chest height!

My preferred hurricane hole would be a small winding creek that was not part of a river system and I would want it lined with a dense stand of mangrove trees on all sides. There would be no habitation for some miles and preferably we would be the only vessel anchored in the creek. On a number of occasions we have sheltered in creeks of this description on the North Queensland Coast whilst gales blew themselves out. It was so tranquil at anchor in these places that it was difficult to believe that the weather outside was behaving so badly.

An ideal bolt-hole from a TRS - up a wooded creek.

**Doldrums Rain Squalls**

To move from the southern hemisphere to the northern hemisphere in the open ocean, or *vice versa*, means that you will need to cross the doldrums areas. As this is the area where the NE trades of the northern hemisphere converge with the SE trades of the southern hemisphere it has become fashionable to refer to it as the Inter-Tropical Convergence Zone, (ITCZ).

These areas consist of a band of air that separates the air masses on either side of the equator. I use the term equator loosely in this context as it is not unknown for the ITCZ to drift some way to the north or south but always remaining within the equatorial area. Air masses are reluctant to mix and those on either side of the equator are no exception. Invariably where two different air masses meet there are weather problems of one kind or another.

The doldrums are best described as areas of virtually no winds liberally sprinkled with extremely vicious squalls. These areas span the equator and extend anything up to 750 miles north and south. To add to the problem for a sailing vessel, the ocean in the Doldrums is a mass of conflicting currents that carry a boat back and fore in a seemingly irrational manner. We progressed a full six miles in the course of our first twenty four hours in the Doldrums. The second day we made forty miles.... backwards! We crossed the Indian Ocean doldrums area from Cocos Keeling to Galle in southwest Sri Lanka, a distance of about 1,200 miles. In the trade winds this would have been an easy ten day trip. It took us forty two days. The ancient mariner knew the score!

Although there is so little wind in the area the sea is constantly moving. The swells are not large by ocean standards but they cause the boat to roll badly enough in the windless conditions to make it necessary to stow the sails to save them from total destruction by chafe. This in turn means that someone must be on watch at all times to seize the chance to make a mile or two of progress when the odd puff of wind appears.

The squalls are something else. They manifest themselves as rapidly building towers of beautiful, snow white cumulus cloud which in its mature form takes on the appearance of the mushroom cloud associated with a nuclear bomb explosion. The stem of the mushroom shaped cloud is in fact the heaviest rain storm you are ever likely to experience. It was necessary for us to enter this area of rain a number of times to collect drinking water.

Before approaching the centre of the storm we would make our preparations; check the sail and dinghy lashings, stow all loose gear, rig the cockpit tonneau and arrange the water collection pipe from the coachroof to conduct the water into our tank. With this done to our satisfaction we would motor into the centre of the squall. As we approached the squall under engine so the wind would rapidly pipe up. Once we were beneath the vast cloud the winds were so strong that we were away like a scalded cat, under bare poles, steering by compass in the hope of making a few miles in the right direction. Within this area the rain would hammer the decks and coachroof with a sound like gunfire. Visibility would be so reduced that we would be hard put to it to see our bowsprit.

Our first water gathering effort of this kind prompted us to use these squalls to make some progress. We would try to assess the direction the squall was moving in and then motor to a position on its outskirts where the wind would permit the use of a little sail. If we got it right we made some progress. Just occasionally we would misjudge the direction the squall was travelling and find ourselves engulfed in it. Then there would be a frantic scramble to get the sails down and secured before the outrageous winds could destroy them. The immense downpour of huge, freezing cold raindrops hammering down would make it both a chilling and a bruising experience. By this means, and by utilising every catspaw of wind that came our way, we made progress, but it was painfully slow progress.

Commonsense suggests that you should fill your boat with diesel and motor your way north. We had been unable to top up our stock of diesel in Cocos Keeling but even if we had we could not have run our engine continuously for a week or more. To live with an engine thudding away for days on end is more than flesh and blood could stand, well our flesh and blood anyway. What is more we would have missed the magic of the Doldrums. Despite its frustrations and the extreme heat in the windless conditions the sea and the sky, particularly later in the day, were beautiful. As the sun dropped lower in the sky the sea would take on the appearance of molten metal of ever changing colours with the sky above trying to outdo the sea for sheer beauty, the like of which we shall never see again.

The wildlife in the sea and the air was a pleasure to watch. We usually had an escort of Dorado, those superb turquoise coloured fish. Ours appeared to be a family, if that is not too fanciful, ranging from about twelve inches to four or five feet long. Dolphin would join us from time to time, sometimes hundreds of them, when the Dorado escort would promptly disappear. Some days large manta rays were seen to jump and could be heard crashing down onto the water's surface. All this and more was ample compensation for the privations we undoubtedly suffered in those forty two searingly hot days.

Snowy white cumulus in the formative stage of a Doldrums rain squall, followed by the torrential rain.

## Tropical Revolving Storms

| Area | Name | Season |
|------|------|--------|
| A-The West Indies | Hurricane | June to November |
| B-The Arabian Sea | Cyclone | Late April to early December |
| C-The Bay of Bengal | Cyclone | Late April to early December |
| D-South China Sea (The worst months are August and September) | Typhoon | Late May to late December |
| E-Northern Pacific Ocean (eastern side) | Hurricane or Cordonazo | June to November (The worst months are August and September) |
| F -South-western Pacific Ocean and Australia's West, North West, North & Queensland Coasts | Cyclone | December to April |

B & C— The Arabian Sea and The Bay of Bengal are subject to monsoon weather. North-easterly winds prevail from early January to late March. From early May to late November south-westerly winds are the norm with some possibility of cyclones.

G— In the South Indian Ocean the south-easterly trades are strong but usually safe from May to November inclusive. From early January to late March south easterly winds prevail with some possibility of cyclones.

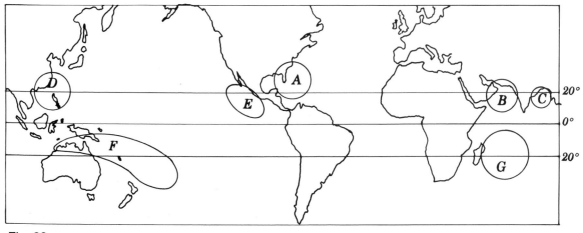

Fig. 29

**The Red Sea**

It helps discussion of wind conditions in the Red Sea if we tackle it in three sections, the total length for part of the year and north of Port Sudan and south of Port Sudan for the rest of the year.

From March to October the wind will be NNW for the full length of the Red Sea. From late July to December the NNW winds north of Port Sudan will be very strong. In July the winds will start to increase and build to a crescendo in October and then start to diminish throughout November and December. When we battered our way north at this time of the year the oil rigs were broadcasting weather reports that regularly included 45 knot winds, gusting to 60 knots and we were in no mood to contradict them. The seas were often three metres high and sometimes higher and so short as to seriously impede our progress northwards.

I have read that the winds in the Red Sea seldom exceed 30 knots. I promise you that they do in the appropriate season, we have the bruises to prove it. The same source also claims that the current seldom exceeds half a knot. I find it difficult to believe that sustained winds of 45 knots and more in a narrow sea will induce no more than a half knot current

South of Port Sudan from about November to April the winds are often from the South but can be variable at times making these the best months to tackle a northerly passage of the southern section of the Red Sea. Sometimes the southerly wind edges into the Straits of Bab el Mendab in the early Autumn, coinciding with the arrival of the NE Monsoon. North of Port Sudan from perhaps November to April the NNW winds will have moderated but the direction will be unchanged.

At the northernmost end of the Gulf of Suez the wind often moderates and may even come in from the south. After an exhausting passage from Jeddah against exceedingly strong head winds we were becalmed, without an engine, in the narrow entrance to the Bay of Suez, possibly one of the busiest shipping spots in the world. Is it any wonder we believe that Murphy's law is over optimistic!

During the strongest winds we found that they moderated considerably close in to the shore. Unfortunately both sides of the Red Sea are encumbered with coral reefs and therefore unnavigable. However it does mean that if you can find shelter in which to anchor, the winds will leave you in peace until you poke your nose out again.

**The Cape of Good Hope**

The passage around the Cape of Good Hope, known to seamen as 'The Cape of Storms', must be tackled with understanding. Yachties I have met who have tackled both Cape Horn and South Africa say that there is little to choose between them.

If the Indian Ocean is to be crossed, the months between May and November give strong SE trades. January to March add the possibility of cyclones, leaving April and December as marginal months.

It is generally agreed that February is the best month to round the Cape and it makes good sense to reach Durban in plenty of time to allow you to celebrate Christmas there. The Durban Yacht Club is renowned as a source of sound advice on tackling the Cape. They have made a practice of offering this advice to passing yachties in the form of lectures around Christmas time.

The difficulties connected with this passage are created by a number of factors. The warm Aghulas Current runs southwest down the coast of South Africa at speeds of one to five knots according to the season. The southern hemisphere summer is the time when the speed of the Aghulas Current is at its slowest. This does much to dictate the choice of time to round The Cape.

A wind that is blowing against a strong current will always kick up more of a sea than one that blows against a more moderate current, but there is more to it than that. Offshore, adjacent to the warm Aghulas Current, are cold currents running in a totally different direction. A warm sea will warm the air above it and a cold sea will cool the air over it. Warm and cold in this context are relative and have nothing to do with the human perception of warm and cold. The meeting of dissimilar air masses is always a source of turbulent weather. The sea behaves in much the same way as the air above it, producing turbulent currents where different waters meet.

Not only does the Aghulas Current slow its pace in the southern hemisphere's summer but the difference in the temperatures of the air masses is smaller, thereby reducing their tendency to antisocial behaviour. A further problem is created by the existence of offshore banks of varying depth that add to the turbulence of the sea. Where the two water masses meet the change of colour and the greatly disturbed sea gives the impression of shoal water and can be a daunting sight.

Waves generated in the Southern Ocean reach the vicinity of South Africa as a swell, having combined with waves from storms etc en route. Naturally such a swell will have a measure of confusion built into its pattern. This swell, meeting the Aghulas Current head on, will shorten and steepen creating complex and sometimes massive seas and when the current is running at its fastest the turbulence can reach destructive proportions.

To combat this mass of doom and gloom it would be as well to keep in mind that a great many yachts make the passage safely each year. Timing is the most important factor. Get to Durban well before Christmas and abide by the advice of the Durban Yacht Club.

Perfect conditions for the daily sun shot.

# Chapter 11
# Damage Control

When a ship is holed below the waterline, collision mats may be used to cover the hole and greatly reduce the ingress of water. A collision mat might be described as a large tarpaulin that is manoeuvred into position with the aid of ropes that go around and under the vessel. Attempts have been made to use the same idea to control hull damage in yachts. Most of the efforts along these lines have failed and the few successful attempts have succeeded largely due to the bloody minded courage and determination of the seamen concerned.

The outstanding example in recent years was Commander Bill King's epic struggle to keep his yacht afloat in the South West Pacific after it had been damaged by a whale. The account is something special and well worth reading. When you look at the photographs you will wonder if the sail used as a collision mat was responsible for keeping the water out or was it the mass of ropes holding it in position that kept the sea at bay. Just about every line that could be spared was used and constant vigilance and adjustment were needed to keep things in place until he reached port.

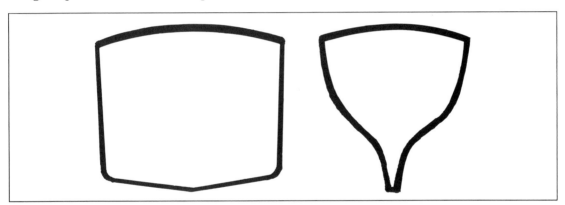

Fig 30. Cross section of a ship's hull on the left, and that of a yacht on the right.

## Below Waterline Repairs

Figure 30 shows two hull cross sections, one is typical of a ship and the other of a yacht. It should be apparent from these two sketches why a collision mat will work for one and not for the other. The Royal Navy claims that it has a drill for everything .... know the drill and when the problem arises your response will be quick and correct. We would do well to adopt the same approach. The drill I was taught for 'hole in bottom of boat' was 'sit the most junior rating in the hole'. It works, I've done it!

Many years ago in a rather nice 26ft clinker-built, centre-board boat called *Avocet*, we were working our way into Leigh-on-Sea just after low water and went aground with our centre board up. No matter, the tide was making. The old fisherman's anchor was taken out and we waited for some more water to arrive. As soon as we were afloat again we motored off and promptly sat on the upraised fluke of the anchor.

I raced below and found the floorboards floating out to meet me. Typical of those days, the forepeak was a sail store and the only furniture was a pair of folding pipe cots. The hole was well forward and I threw the clutter of sail bags, ropes and so on out of the way and sat in the hole. Magic! no more water. So long as I remained in position we were safe. A considerate skipper gave me a kapok cushion to sit on and I was no longer in danger of developing gravel rash on my posterior as we bumped across the sandy shallows.

To see daylight and a great gout of water issuing from a six inch diameter hole in the bottom of a boat was scary, and on that occasion we could have walked ashore. I am sure that had my reaction not been instinctive the shock would have slowed up my response considerably.

When the boat was bailed dry it was decided to make some sort of repair. We, or rather they, as I was still on duty, found a board that would span the hole and overlap the ribs on either side of it. A handful of nails and three or four towels completed the repair kit. The pad of folded towels was exchanged for my buttock and the board was nailed firmly in place over the pad.

This, admittedly limited experience, convinced me that collision mats had to give way to something stuffed into the hole from inside the boat. If you accept that, it must follow that it should be possible to reach every part of the hull from inside the boat. I agree that this is a counsel of perfection but it is the ideal to work towards.

We have always carried a small crowbar and a three pound club hammer to allow us to tear away any interior construction that would stop us reaching a damaged area. Finesse has no place in your immediate action ....get to the hole and plug it with something .... anything. Then you can make a cup of tea and sit down and think of a more permanent solution. In our emergency kit we carry a number of folding wedges. If you are not familiar with their use you should look around boat yards and building sites. Two wedges placed under the end of a strut and driven together with a hammer can exert tremendous pressure. I can visualise the towels, nails and board technique we used in *Avocet* being modified with struts and folding wedges to work inside a glass fibre boat.

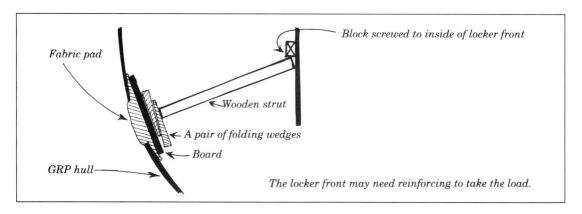

Fig 31. The use of wedges and a fabric pad to block a hole in the hull.

Later, if you can dry your boat out on a beach, it should be possible to apply a temporary patch with some thin plywood if the damage is not too extensive. A padded plywood patch both inside and outside, liberally anointed with mastic and bolted together through the sound fibre glass around the hole seems likely to work. Carry some two pot resin glue that is intended to set under water. You may find it as useful as we did.

After taking us almost all the way round the world, our self-steering gear was partially wrenched off in a vicious squall off north-east Crete. This left four ³/₄ in diameter holes just above the water line and our underwater resin glue made a perfectly good repair.

## Window and Portlight Repair

Windows can be stoved in by the odd rogue wave that runs counter to the general pattern of seas in which you are sailing. Such a wave can strike with a disturbing force and it makes sense to carry a quick repair kit. Such a kit would consist of a stout piece of ply, two battens, two nuts and bolts and four washers.

The ply and the two battens of 2in x 1in soft wood should have matching pairs of holes to accept the bolts. In use the ply would be placed to cover the window on the outside. With the two battens spanning the inside of the window the whole could then be bolted together.

## Rudder Damage

Rudder damage can amount to loss of the rudder itself or a failure somewhere between the wheel and the rudder. It is likely that the axle of your wheel fits into the centre of a large cog or sheave. The axle and the cog or sheave may be held together by a steel pin that passes through both of them. This should be a high tensile steel bolt (car spares shop) and you really should include a couple of spare bolts in your stores.

Your wheel may have a key and keyway to unite the parts, instead of a pin, and this is a much better engineering answer and is likely to last a very long time. If your wheel is connected to your rudder yoke by steel wires, you should carry a few bulldog grips of a suitable size to enable you to make a repair should a cable break. On the other hand if you are relying on a hydraulic system to turn your rudder the problem may well be insoluble at sea and you will wish that you had something simpler.

Our rudder seized solid about 200 miles from Portsmouth, for which we were making at the time. An axle and a wheel had been fitted with so little clearance that it was impossible to introduce adequate lubrication. We have since had the parts machined to increase the tolerance between them and installed a grease nipple to let us introduce grease where it is needed.

At the time we fixed a short length of steel piping to the quadrant that sits on the top of our rudder stock to act as a tiller. Lines were attached to this tiller and brought to the cockpit through a series of blocks so that we could steer with tiller lines. This system worked but it was very tough on the hands. After a four hour watch, assistance was needed to uncurl one's fingers from the ropes. If you find yourself confronted with this problem some more comfortable hand holds than one inch rope would need to be devised. I had designed our self-steering rudder so that it could be used to steer by hand but of course Murphy had seen to it that I buried the self-steering rudder at sea after it sustained storm damage off N.E. Crete.

You may feel it to be worthwhile to drill a hole about 1¹/₄ in diameter, near the back and top of your rudder. The hole can be filled with putty when it is not needed. When your steering control fails the putty could be replaced with a rope that has a stopper knot either side of the rudder. With the ends of the line brought up to the cockpit some measure of control could be devised. If you have no hole in your rudder blade take the pin out of a large shackle and secure a line to either side of the shackle. Jam the mouth of the shackle over the after edge of the rudder and bring the lines to the cockpit.

The ship's wheel speaks for itself.
Part of our emergency steering control can be seen running through the double blocks.

Should you lose a rudder then a whisker pole must be sacrificed. Bolt or lash a board to the end of the pole and rig it over the stern like a sculling oar, lashing it securely where it meets the deck or guard-rail. Because the jury rudder will be further aft than the ship's rudder it will exert greater leverage and so it is probable that a smaller, less efficient blade will serve. I know of a yacht that crossed a large part of the Indian Ocean with a jury rudder that consisted of a spar and a board over the stern, but in this case the spar was fixed rigidly on the fore and aft line so that no normal rudder movement was possible. The only movement was a rotation of the spar so as to turn the blade from its vertical position. By rotating the blade one way or the other, it is claimed, it was possible to steer a reasonable course.

This is the system that I think I would try first as it is claimed that it has worked and if it does, it satisfies my overwhelming desire for simplicity. If it fails,, you will not have wasted your time, the blade and pole could well be the basis for another approach.

**Broken Main Boom**
A broken boom is certainly a nuisance but it is not a disaster. When you have cleared up the mess and salvaged all you can for future use the first step will be to get under way again. A whisker pole may perhaps be called into service to pole the main out or it may be that your main will set reasonably as a loose footed sail. Failing both these possibilities you might find that your storm trysail will be more amenable to being set loose footed.

The important thing about breaking a boom is to decide why and how it happened so that you don't do it again. I speak with feeling on this subject as we broke two booms before I realised why it was happening. On both occasions we were deep reefed. In this state the clew of our main comes about halfway along the eighteen foot long boom. With one end of this very long boom restrained by the sheet and the other by the gooseneck, the reefed main put too much load on the unsupported middle section of the boom with unfortunate results.

The mainsheet lead has been changed so that one block is now fixed at about eight feet from the outboard end of the boom giving additional support where it is needed. We have sailed about twenty thousand miles with this new sheet arrangement and have had no further problems.

*Didycoy*'s original boom was made of timber, so there was no thought of going to aluminium for its replacement. The original spar was a solid wooden pole with a diameter of about three inches. Its successors were both laminated from one inch boards with a rectangular cross section of about 3 x 4in. I'm not at all sure that lamination was needed, the faulty sheet arrangement was so obviously to blame. Certainly I would not hesitate to use a solid spar if suitable glue or facilities were not on hand.

Because the boom that has broken was made of aluminium, it does not follow that it's replacement must also be made of the same material. You will only be able to replace an alloy spar in relatively sophisticated places and then it will be at great expense. Part of this expense will be the cost of air freighting the new boom from far away manufacturers which will probably cost as much as the boom itself, what with insurance, freight agent's fees, forwarding agents, clearing agents, customs, bonded storage, bonded transport and quite likely schmeer geld to unstick the boom from a multitude of sticky fingers. Timber is available almost everywhere at a fraction of the price of an imported alloy boom. The boom is unlikely to damage the goose neck when it breaks so it should be possible to utilise the existing fitting. Later, when you wish to sell the boat, will be the time to buy an expensive new aluminium boom to improve the resale value of your yacht.

If the boom has bent at about the halfway point it should be possible to saw out the damaged section and then drive in a length of shaped timber that will act as an internal splint. Most of the fittings will be on the ends of the original boom and so much work is likely to be saved by uniting the two pieces by this means. I'd be tempted to put a seizing around both cut ends of the damaged boom where it meets the internal splint as this looks like a weak spot and I would screw both aluminium parts to the internal splint because an involuntary gybe can apply quite a backward pull on a boom for a short time.

I think both ends of the plug should be slightly tapered over several inches. It seems to me that if the end of the plug is tight fitting and square to the axis of the boom it will subject the aluminium tube to undue stress where the two meet. Hopefully a very slight taper will spread the load a little.

When a boom is broken it is likely that the track for the foot of the mainsail will be damaged or your repair will make it impossible to use the track. There is no reason why the foot of the main should not be laced to the spar, this will serve just as well as a track.

**Broken Mast**

Unlike a broken boom, a broken mast could qualify as a serious problem. Obviously you would salvage all you could for future use and my imagination suggests that will not be easy. Most passage making vessels seem to be manned by a crew of two, usually a man and a woman and that is not a lot of muscle power with which to retrieve a mast that has gone over the side. Nevertheless make the effort, salvage all you can.

The oft repeated advice that you should not start your engine until you are quite sure that every rope and wire is either aboard or a very long way from you, bears repeating. We are creatures of a technological age and I suspect that the sound of the engine performing well in a crisis is a great comfort to many people, so keep your finger off the starter button.

Sailing on the ocean is not like sailing near the coast, in narrow channels or amongst other traffic. With no other vessels to worry about and the nearest land many miles away you can afford to take the immediate action and then sit down and have a drink or a bite to eat while you think things over. If you are in shoal water, drop an anchor to give you time to clear up and sort things out.

If you can rig a jury mast with one or more pieces of the damaged mast you will be back in business. Don't forget that you have winches around the boat, including the anchor winch, that will help you recover some of the gear that has gone over the side.

With a stump of a mast it might be possible to turn one of your sails into a lateen sail to gain some height by using whatever spars you have on hand. Dhows are lateen rigged and they sail well, but the rig is not one that encourages easy manoeuvring in tight circumstances. All you want is a rig that will get you within easy motoring distance of safety.

A dhow's mast has one set of shrouds that are rigged on the windward side of the mast. When they are about to tack they raise the spar to the vertical to allow it to pass around the front of the mast. The mainsheet is brought forward of the mast and along what is to be the new leeward side to its normal place near the helmsman. Meanwhile others are working to free the shrouds and take them to the new weather side where they are secured. That's fine in a dhow with a dozen or more crew but in the average yacht it would take a little longer!

'A lot of work'. do I hear you say? Yes, you are probably right but it serves you right for using inadequate standing rigging.

Looking out from under our self-made main boom towards an anchorage near the Great Barrier Reef.

# Chapter 12
# Liferaft or Lifeboat?

It is generally agreed amongst passage makers that a liferaft is of little value to them. There is no point in sitting in a slowly disintegrating liferaft waiting to be picked up by a passing ship in an area where there is no shipping. When a ship is sailing far out on the ocean their look-out is understandably nothing like so keen as he would be in the North Sea. There have been numerous cases reported of castaways not being seen even though they were firing off rockets and flares within sight of a ship. Although we were never castaways, our own experience suggests that this scenario is what we would expect in view of our apparent invisibility to a large number of ships.

If a vessel is aware of your need you will be treated with a generosity and kindness that will likely surprise you. The problem as in so many other spheres is communication. There is no question in my mind that once you are away from home waters you must behave as if you are on your own, which will likely be the case whether you believe it or not.

What is needed in the event of disaster is a buoyant, seaworthy, mobile lifeboat so that you can make your own way to safety. The very act of sailing your own lifeboat towards safety is a positive step towards survival. To sit helplessly in a liferaft waiting for someone by chance to come your way, *and to see you*, must be totally destructive of morale. Two things must be clearly understood. The only people who survive are those who are determined to do so and work at it. Those who sit and wait and hope, are doomed from the start. The other point you must understand is the near total lack of shipping in the areas in which you are likely to be sailing. There will be no one to pick you up.

Liferaft service stations are few and far between in blue-water sailing areas and I suspect that constant exposure to the sea and tropical temperatures is likely to cause a liferaft to require more, not less, servicing. In tropical waters, the need for the protection given by a liferaft's canopy is nothing like so great as the need for shelter in colder latitudes. It is true that the sun can sometimes be a problem but unless you are very unlucky you will have developed a degree of tolerance before you need to take to your lifeboat!

One of the many things that bother me about a liferaft is that you have no idea what is going to come out of that container until you have pulled the cord; will it float, will it inflate, will it stay in one piece once it has inflated? The one thing you can be sure about is that it will not take you anywhere. Don't be seduced by the word 'liferaft' it could just as easily be a death trap.

The 1979 Fastnet race storm saw numerous examples of liferaft failure. Seven deaths occurred amongst crews who took to their liferafts in the course of this storm. Ironically the boats that were abandoned in these incidents were later recovered and towed in. In all seventeen boats were abandoned and all were recovered. This must suggest most emphatically that the safety that is represented by a boat, even though damaged, should not be abandoned until it is seen to be disappearing beneath your feet despite your best efforts. I have personally helped to work on a twenty five foot boat that had water over the bunks in the main saloon and although she felt like a half tide rock, she stayed afloat and we were able to pump her dry.

In the RYA report on the Fastnet storm one crew is reported to have inflated their raft 'because water was coming in faster than they could expel it.' The inflated raft blew away.

This so concentrated their minds that they returned to bailing with renewed enthusiasm and had a dry boat in less than two hours! I have always wanted to know what the skipper said to the crewman who tied that liferaft to the guard-rail.

In conversations with potential passage makers it has become obvious that most people believe that if they have to abandon their boat it will be in the course of really bad weather. If you have chosen a seaworthy vessel this is the least likely time for such a disaster to strike. The possibilities, as I see them are, a fire that gets out of control or collision with a floating or stationary object. Your charts will show you where the stationary objects are, collision with a floating object at five knots is not likely to do serious damage to a proper sea-going boat, unless of course it is a ship, and the number of really aggressive whales in the world is minuscule. As for fire, a proper understanding of the precautions to take will go far to eliminate this as a likely event.

## Inflatable Dinghies

Your choice of tender will lie between a hard dinghy, a folding dinghy or an inflatable. If you wish to dive or snorkel you will need an inflatable. The average hard dinghy will capsize if you attempt to climb into it from the water. (If you must regain the safety of a hard dinghy go in over the stern, it is probably the only hope you have of getting aboard without capsizing it.) The doughnut shaped inflatable is the least satisfactory design by far. An outboard is an almost essential item if you are to choose an inflatable. The rowing qualities of the average inflatable are abysmal. The oars provided are too small to permit you to row the beast in anything but a following wind. When you substitute a man-sized pair of oars you very soon discover why the manufacturers provide a pair of toy oars; the rowlocks pull off the buoyancy tubes!

When we were in the Galapagos I saw a dodge that improved the rowing qualities of an inflatable to a remarkable degree. A plywood board about two feet square had been attached to an outboard bracket as a sort of fixed rudder, or perhaps an outboard centre board would be a better description. The outboard bracket allowed the centre board to be lifted as shoal water was approached. We were in the open anchorage of Academy Bay when we saw this arrangement and the wind and sea were of a kind that would have made rowing an unmodified inflatable at least difficult and maybe impossible but it was moving well. When I chatted with the owner about it, he claimed that it had transformed the handling qualities of his inflatable. He found that it paid to have two people to paddle rather than try to row it.

An inflatable is closer to a liferaft than are hard dinghies with the added advantage that every time you use your inflatable you are aware of its condition. You only discover the true state of your liferaft when you inflate it and then it will be too late to effect repairs if they are needed. It is possible to have valves fitted to an inflatable that will permit emergency inflation with $CO_2$ and this would be a sensible move. When deflated an inflatable stows well and if space is limited this may be the overiding argument in any discussion of its merits.

The inflatable must be made mobile somehow if it is to serve as a lifeboat. There is a Finnish sail kit that clamps on the transom of a dinghy in much the same way as an outboard motor. A mast and sail are mounted on the outboard bracket with an outsize rudder hung below. The oversize rudder also acts as the centre board. It certainly works well on a hard dinghy and I would think it likely that it will do the same for an inflatable. After all, you will only need to run down wind or at the most across it.

The photographs of Desmond's fourteen foot tinny shows this system in action. Tinny is the Australian name for a can of beer, but when a tinny reaches these dimensions it is usually a pressed aluminium or steel dinghy that is referred to. Desmond had sailed his tinny about 800 miles up the east coast of Australia when we met him. He is a larger than life character who claimed to know very little about sailing and proved it by the manner of his coming alongside *Didycoy*. The paint I replaced the next day, but the scar needed a tablespoonful of filler before it healed.

A cabin trunk, a large suitcase and a kitbag or two held his possessions. Each night he would run his boat ashore and set up camp on some deserted beach or cove. He was having a wonderful time. His foremast and sail came from an old sail board but the after mast was a Finnish transom-hung sail and rudder. He was able to sail on a reach but I suspect that a more skilful hand could have done better than that.

If an inflatable has a hard floor it should be a simple matter to install a square sail, a lateen or a sprit sail right forward and hang a rudder on the transom. This would allow you to run before the wind and if you sail before the trade wind for long enough you must meet up with land sometime!

There is a selection of collapsible dinghies on the market at present and many of them look good. If space is at a premium this may be an area worth investigating.

Desmond's Tinny, seen in action with the Finnish outboard sail drive unit. This unit has great potential for use on an inflatable lifeboat.

101

## Hard and Folding Dinghies

The comments and suggestions concerning the mobility required of an inflatable apply equally to a hard dinghy or a folding dinghy. In addition the hard/folding dinghy must have sufficient buoyancy to keep it afloat in all circumstances. This may take the form of built-in buoyancy or buoyancy can be added. If buoyancy is to be added I would favour inflatable tubes that are designed to be attached to the outside of the dinghy. In addition to adequate buoyancy they give a degree of stability that is only to be found in an inflatable. Buoyancy tubes of this kind should be fitted with valves to permit rapid inflation with $CO_2$ in an emergency.

As well as being your lifeboat and your link with the shore, your dinghy will, at times, be required to carry considerable loads of diesel, water and shopping and to enable you to lay a second anchor. And finally, a good dinghy can enhance your pleasure by allowing you to explore nearby beaches, reefs and creeks that might otherwise be inaccessible to you.

Should you have the misfortune to overturn a hard dinghy you must first turn it right side up. There will be no point in trying to climb into the dinghy until you have removed a large part of the water. The quickest way to do this is to bear down heavily on the top of the transom and then give a powerful forward thrust. This should expel enough water to let you climb in over the stern and bail out the remaining water .... if you had the foresight to tie your bailer to your dinghy.

## The 'Going Away Bag'

A 'going away bag' is obviously an essential item should you need to abandon ship in a hurry. Here are some thoughts on what should go inside.

Thermal blankets, designed for mountaineers and others, are minute when folded and have remarkable heat retaining properties. It can get cold at nights even in the tropics. They are also waterproof. We put a few brass eyelets around the edge of ours so that we could secure it inboard if need be. Not only will these blankets keep you warm but they can serve as a rain catchment, an awning or a sail.

One gallon bottles of water will be easier to handle and stow in the dinghy than larger bottles. Don't completely fill them and then they will float in sea water. Link several bottles together with a light line and secure one end to your dinghy before you launch it. These two measures will let you throw the bottles over the side and forget them until you are settled in your lifeboat. Use plastic bottles that are strongly coloured to exclude sunlight otherwise your water will develop algae in a very short space of time.

It is strongly recommended that castaways should drink nothing for the first 24 hours to reduce excretion. Service survival courses advise that up to 40 per cent of sea water can be added to fresh water on hot days when the subject is perspiring freely. In the small boat situation perspiration will evaporate as soon as it appears and you probably wont see it. I must emphasis that this is an occasional measure, not a daily one, since too much salt taken into the body can be damaging. It is only possible to eke out your supply of fresh water in this way because excessive perspiration does drain the body of salt.

Frequent dousing of the head and body with sea water will help to fend off dehydration simply by keeping you cool. The head is very well supplied with blood vessels and it would pay to wear a hat simply because the proliferation of blood vessels will absorb more heat than other less well provided areas of the body; more heat, more perspiration, more dehydration.

If you are desperate for water and close to a doldrums area, it might pay to make the effort to enter the area with the intention of collecting some water. The squalls in the doldrums consist of immensely strong winds and tremendous quantities of rain. Having experienced these squalls at first hand I would say that this is an heroic solution but if the alternative is to die of thirst .....

Some light monofilament and small fish hooks could be of value. If the flesh of a freshly caught fish is diced and placed into a handkerchief which is then twisted to expel the fluid from the flesh it can be drunk as a supplement to your drinking water. The eyeballs are full of edible fluid, don't waste them. I have tasted better flavoured drinks but if you are thirsty enough you will be glad to have it.

The digestion of protein requires significantly more water than does the digestion of carbohydrate so this should guide you when you are packing some food for the bag. If you intend using a rubber dinghy as your lifeboat it might be wise to include a cutting board!

A small mirror seems to have served a number of castaways well as a daylight signalling device when used to reflect sunlight into the bridge of a distant ship. Certainly it was the only thing we found that attracted their attention when we desperately needed it.

If your continued health, or maybe your survival, depends on a daily dose of a particular drug then make sure you have at least three months supply in a waterproof container.

It has been said before but it merits repeating. The most important factor to take with you is a bloody minded determination to survive. Above all else, do not abandon ship until despite all you have done, she sinks beneath you.

A Red Sea passenger.

# Chapter 13
# Fire Fighting

When I step aboard another vessel I can usually tell at once if they have ever experienced a fire whilst afloat. If they have they will be as richly endowed with fire fighting gear as is *Didycoy*. I shudder at times when I see how poorly equipped some boats are. Indeed, I think that some of the gear I have seen has been so small that it can be of little more value than a lucky charm.

Before we experienced our own fire we had three $1^1/_2$ kg extinguishers, a small fire blanket and a bucket on a lanyard. You would perhaps be forgiven if you thought this to be a reasonable outfit, it is certainly more than many boats carry. We now have one 4kg and three $1^1/_2$kg dry powder, one BCF, and two CO2 extinguishers, a fire blanket and a galvanised bucket on a lanyard. In addition we also have two feet of soft plastic tubing that will slip onto the electric pump-operated fresh water tap or the salt water tap that is operated by a foot pump. Most fires start small and quickly grow if they are not tackled at once and this is where I think the galley tap hose could be of value if it could be brought to bear.

We used our three extinguishers on our very own blaze and I assure you that to look your last extinguisher in the eye with the fire still burning brightly is a sobering experience. We were lying in Middleburg in Southern Holland and I had a charter party of five men aboard. During the day I had repaired a leak in our kerosene cooker but unknown to me a quantity of kerosene had soaked into the mineral wool insulation surrounding the oven. When we had berthed the boat for the night I sent the crew ashore to the Arne Yacht Club to slake their thirst while I cooked the dinner.

Within minutes of lighting the oven I had a solid rectangle of flame that reached from the cooker to the deckhead. I shut off the supply of kerosene and put the flames out with a blast of BCF gas from the extinguisher. Almost at once the flames were back again. The cooker was now so hot that each time I put out the fire the flames were back again leaving me insufficient time to get the bucket from the cockpit locker just six feet away, let alone fill it with water. I carried on like this for one and a half extinguishers. The heat was increasing and I was being forced out of the cabin. I think I was the luckiest person alive on that occasion. Can you believe that amongst the assembled boats' crews were two professional firemen, one Dutch and one English, who came to my assistance.

When it was all over I realised that there were about one hundred spectators, some actually leaning on the nearby drinking water hose pipe that would easily have reached us. Not one of the bystanders lifted a finger to help — not even to call the fire service from the nearby telephone. It is no exaggeration to say that we would have lost *Didycoy* if it had not been for the arrival of those two guys. I shall always be grateful to them.

Naturally next morning I questioned them as to what was wrong with my technique. The first thing they said was that if *Didycoy* had been built of glass fibre instead of wood, a fire of that size would have quickly been out of control. This is the essence of what they told me:—

1. Lie on the floor to fight a fire in a boat. If you insist on standing your head will be in the hot smoke-filled air. At floor level there is fresh, cool air.
2. Do not fire off an extinguisher at a cloud of smoke, it will probably be wasted. Find the flame and hit the base of it with the extinguishent.

3. Be prepared for the fire to reignite, maybe repeatedly.

4. A bucket of water will not only extinguish a fire but will also cool the area so that it will not reignite.

5. If the burning material is safely and easily portable throw it over the side.

That is a fair set of general rules but of course there are specific rules too.

1. If applicable turn off the fuel supply.

2. Do not throw water onto an oil fire, it will cause it to spread violently.

3. Do not use an extinguisher on a pan of cooking oil that has ignited, it will splatter burning oil around the boat causing more fires to break out. Do cover the pan with a lid of some kind, thereby denying it air. Turn off the cooker and leave it until it has cooled enough not to reignite. On no account carry a pan of blazing fat around the boat, perhaps with the intention of throwing it over the side, you are bound to spill it, probably on yourself with possibly disastrous results.

4. Don't use water on an electrical fire. Switch off the main and if the fire persists use a gas or dry powder extinguisher.

5. Dry powder extinguishers are good but the powder cannot get around an obstruction, it must strike the base of the flames. If the source of the blaze is not directly accessible you must use gas to blanket the area. The vibration created by sailing and motoring will cause dry powder to consolidate inside the extinguisher. To overcome this occasionally invert the extinguisher and shake it vigorously.

---

**REMEMBER:-**
INSTANT ACTION
DENY THE FIRE AIR
COOL THE SITE DOWN
FIGHT IT FROM THE FLOOR
HIT THE BASE OF THE FLAMES
BEST OF ALL DON'T HAVE A FIRE

---

# Chapter 14
# Medical Matters

There are several areas here where you will need to enlist the aid of your doctor. Understandably, doctors generally view DIY medicine with disfavour and there are four things that you can do to overcome this reluctance. First you must know what you are talking about and what you want from him. Secondly you will need to convince him of the near impossibility of summoning help on the ocean. Once away from the land, if you see two or three ships in a month you will be doing better than the average yacht.

Next he must be made to understand the length of time it will take to cross something like the Atlantic ... shall we say 30 days as opposed to a mere six hours by plane? Then finally, and most difficult of all, you will need to convince him of the utter futility of turning back and trying to beat back to your departure point. Perhaps I can give you a little ammunition on that score.

We crossed from the Galapagos to the Marquesas with the trade winds behind us and it took us twenty seven days. At about the same time a yacht tried to make it against the trades in the opposite direction. They gave up after one hundred and forty two days and returned to the Marquesas.

We battled our way north in the Red Sea against very strong head winds and made good three hundred and fifty miles in fourteen days even though we were doing about one hundred and twenty miles through the water every day. In other words we sailed some one thousand seven hundred miles in order to make three hundred and fifty in the right direction. Our engine was of no use to us in winds of the strength we were experiencing, but at that time it was our only means of recharging the batteries. When the engine stopped working we turned back and ran down wind three hundred and fifty miles to Jeddah in Saudi Arabia. It took us just three days.

## Inoculation

Consult your doctor about the need for injections to protect you against things like tetanus, polio, cholera, typhoid and whatever else he deems would be sensible. Don't leave them until the last minute. A second dose of something may be necessary and you must give your doctor time to do the job properly.

Unless you intend to tramp the forests of Panama and one or two other similar areas, yellow fever is not a problem. We were both inoculated against yellow fever before we left and have the certificates to prove it. We had been told that proof of inoculation would be needed in some places and especially for the Panama Canal. No country we visited asked if we had been inoculated against yellow fever and nowhere were we aware of a danger of it. Should you, for some reason, wish to be protected against yellow fever your doctor cannot help you. Stocks of the vaccine are held at some of the larger airports and at the London School of Hygiene and Tropical Medicine and it is to one of those places that you must go.

The effect of some inoculations is short lived and so there may be no point in receiving them. Obviously your doctor is the person to consult in these matters. There are other things that you will need to discuss with him and they will come to light as you work your way through this chapter.

**Possible Common Illness**
The possibility of developing appendicitis is rightly something that should concern prospective passage makers. In the past some folk have actually had their appendix removed before they set out, so serious is an attack of appendicitis, far from medical help. Certainly if your appendix becomes inflamed and ruptures, as they will if not treated in time, you are on the way to becoming the centre of interest in a burial at sea ceremony.

The correct antibiotic will give you an excellent chance of holding it at bay until you can reach medical assistance. The same sort of treatment can keep a tooth abscess at bay. I must underline that all the treatment can do is to hold the infection at bay, not cure it. The cure must be carried out at a hospital ashore.

Discuss this with your G.P. and don't forget to ask him to give you the signs and symptoms of appendicitis while you are about it; its no use knowing how to treat it if you can't recognise it!

It is this idea of what might be called extended first-aid, where your doctor could be so helpful with prescriptions for drugs and so on if you can win his co-operation. Some knowledge of first-aid is important, but at sea the situation does not stop there. Your patient will be in need of care until he recovers or until you reach medical assistance. What you do after the initial first-aid treatment has been administered is just as important as your immediate action.

**Burns and Scalds**
I would also ask his advice on the treatment of burns and scalds, the same thing really, just different causes. Today there are spray on treatments that are a great advance on a swab soaked in cold tea that was my Granny's favourite answer. Immediate immersion in cold water will help to reduce the temperature of the burned area and so limit the damage to the tissue. If a garment is soaked in boiling water or hot fat get it off as fast as you can for the same reason.

Serious burns create considerable shock, loss of fluid because the skin is no longer there to keep it in, the very real possibility of infection because the skin is no longer there to keep it out and a great deal of pain because the nerve endings are exposed by the loss of skin. The problems of fluid loss, infection and pain can obviously be reduced if some sort of skin substitute can be provided and this seems to be the way today's treatment goes. There are spray on treatments for minor burns which may be all that is needed once the area has been cooled down. Most of the sprays are soluble in water and so must be renewed if the area is made wet. There are spray on treatments for more serious burns and it is probable that your doctor can help you with this. If the spray on treatment is successful it will help reduce fluid loss, keep infection out and considerably reduce the pain.

Another recent introduction for the treatment of more serious burns is *sterile, non-adhesive* burn dressing. They are available at the larger pharmaceutical stores and it would make sense to lay in a stock of them. It is imperative that the burn dressings remain sterile and later in this chapter there is a section that will help you understand how to handle them to avoid contamination.

Do not apply grease in any form to a burn that is to go to hospital. The surgeon will need to remove the grease before treatment can be started and this will cause unnecessary damage.

If the burn becomes infected, an antibiotic will be needed and again it is your doctor's advice that you want, nowadays there are antibiotics tailored for specific needs. If the burn is extensive it might even make sense to start a course of antibiotics as a precautionary measure.

Fluid loss should be countered by giving the patient plenty to drink. If you have some rehydrating fluid ( more about this later) it would make good sense to include some in the drinks that you give. Alcohol must not be given, it can only make matters worse.

## Shock

Shock will be present to a greater or lesser degree, not only with burns, but with any injury. The simplest way to describe the effect of shock is for you to imagine that all the blood vessels have become rather slack and so there is no longer enough blood to fill them properly. If the patient insists on standing up it wont be for long because gravity will drain much of the blood from his head and he will become unconscious. Far better for the patient to lie down and for you to raise his feet a little. His feet can manage with a diminished blood supply far more readily than can his brain. It also helps to keep the patient warm. Should the patient become unconscious there are special steps that you must take to safeguard him, these steps are discussed later in this chapter.

Before I continue with this depressing theme let me say loud and clear that I have never ever before seen such a healthy, undamaged bunch of human beings as your average group of passage makers, so let's not despair. It is like so much in this game. One prepares for the worst and it seldom materialises.

## Cuts and Bleeding

Most forms of external bleeding will respond to the same simple form of treatment. Direct pressure with a finger or thumb for long enough to allow the wound to clot will stop most bleeding. After all, blood vessels are only tubes and if you squash the open end with firm pressure the blood can't come out. If it cannot come out, given time, natural forces will cause the blood around the wound to clot and seal the cut.

So often, for some unknown reason, a bleeding wound is held under running water. This can only result in continued bleeding because the clotting blood is constantly washed away before it can do its job. A big fat thumb pressed down on a leaking blood vessel gives a dramatically satisfying result. When the bleeding has stopped some antiseptic and a firm dressing should be applied. If the wound is extensive it may be necessary to substitute a large pad of gauze for the thumb and to hold it in place with a firm bandage

Whatever you do, do not try to stop the bleeding by tying the bandage tightly enough around the limb to cut off the flow of blood, that could be disastrous. A ligature tied as tightly as this could behave like a tourniquet and tourniquets are only required if the bleeding is from an artery and even then, they must be used with discretion. After a maximum of twenty minutes a tourniquet must be eased to allow blood to flow through the limb even though this may mean loss of blood. To fail to allow blood to flow to the limb after twenty minutes have elapsed will cause gangrene to set in and that means a death sentence in our situation.

Arteries are the large elastic vessels that help to pump blood from the heart and distribute it around the body. If an artery is severed it is a serious matter because the blood is pumped out under considerable pressure and therefore the blood loss ranges from serious to disastrous if it continues to flow unchecked. Fortunately the arteries, with a few exceptions, are deep seated and therefore well protected for most of their length.

To sever a major artery requires an accident of the kind that it is difficult to believe could ever happen in a small boat. This is fortunate because whilst the initial first-aid is straightforward and easy to put into practice, the extended first-aid ranges from doubtful to hopeless.

The hands would seem to be the likeliest area at risk, perhaps from a glass that is broken in the course of washing or drying up. Because the hands are at the end of the artery that runs from the heart to the fingers the artery will be at its smallest and this gives hope for a successful outcome. Direct pressure followed by a sterile pad and bandage should be enough to control the bleeding and give nature a chance to heal the wound. If the wound is in the palm of the hand a large sterile pad the size of a current bun placed on the wound, the fingers closed onto the pad and the fist firmly bandaged to apply modest pressure to the wound should suffice. A sling that raises the hand above the heart will help to stop the bleeding and give a measure of comfort.

A gaping wound, large or small, must have its edges pulled together to help it heal. A doctor would likely use stitches to do the job. We should attempt to pull the wound together with a dressing and sticking plaster and think of stitching only as a last resort, more about stitching later. There are things called butterfly sutures that are really shaped sticking plasters and these are designed for this job.

If the wound is to the back of a hand, once the bleeding is controlled and a dressing is secured, it might pay to strap a small padded splint to the underside of the hand and wrist. This will stop the wrist bending and thereby avoid the wound being pulled open.

## Broken Bones

Why broken bones are not commonplace in passage making boats I cannot think but it is a fact. I can recall just one case of a broken limb amongst the considerable number of blue-water sailors I have met or read about that actually occurred at sea. If you or yours are going to supply another exception then you are going to have a problem.

The first-aid advice is to immobilise the limb and then seek medical assistance as soon as possible. This is sound advice but the latter half of it could prove rather difficult to follow if you are ten days from anywhere. Certainly immobilise and support the limb at once with a splint that is padded to accommodate itself to the curves of the limb it is supporting. Inflatable splints are on sale and could be a worthwhile purchase. They have the advantage that they mould themselves to the shape of the limb as they are inflated.

First-aid books will illustrate ways and means of immobilising various broken bones. If the damage is to a shoulder, an arm or the ribs, a sling to take the weight of the full length of the fore arm will give a large measure of comfort. After that it is a question of treat for shock, pain-killers as needed and post haste for the nearest source of help. Healthy people don't usually die from an immobilised broken bone. Nature soon takes over and starts the repair process and the bone begins to knit together. Unfortunately the broken bone may be displaced with the result that the surgeon may need to break the bone again as a first step towards restoring proper function to the limb.

Whatever you do in the way of first-aid do not attempt to straighten a broken bone, thinking to help the situation outlined above. You will almost certainly cause further injury, possibly of a very serious nature.

If you are faced with a compound fracture, one where there is a wound at the site of the break which could allow infection to enter and maybe reach the bone, as well as splinting, some antiseptic and a sterile pad and bandage should be applied. In addition a course of antibiotics must be started.

The area around a broken bone will often swell and this can cause pain. Raise the injured limb a little to let gravity help drain away the fluids that are creating the swelling. Check that the bandages that are holding the splint in place are not too tight and maybe adding to the swelling.

109

Before you can treat a fracture or a break, both words mean the same thing, you must be able to recognise it. The initial letters of the signs and symptoms of a fracture can be put together to form 'slip duc'. 'How did you break your leg? Did you slip duc?' Sorry about that!

'S' swelling. 'L' loss of power. 'I' irregularity. 'P' pain.

'D' deformity. 'U' unnatural mobility. 'C' crepitus.

Irregularity may be detected if a finger is run lightly along the bone in the area of the injury. You must never test for unnatural mobility but it may be apparent. Crepitus is the grating noise that might be heard if the broken ends of the bone are rubbed across each other. Never test for crepitus. It is only included because the patient may have heard or felt the effect. If swelling, loss of power and pain are the only symptoms present the chances are that you are not dealing with a fracture or if you are it is hopefully not a particularly bad one. In any case the treatment for a sprain, strain or fracture is the same, support, rest and immobility of the injured part.

The extended first-aid treatment must be aimed at keeping the patient as comfortable as possible subject to the need to keep the injured part immobilised. Treat for shock if it is present and administer pain killers as needed. Get to medical assistance as soon as possible.

After a limb has been splinted, check at intervals to see if blood is reaching the fingers or toes as the case may be. This is easily done by pressing the finger or toe nail quite firmly between thumb and finger. If the blood supply is as it should be the pressure on the nail will cause it to lose some of its colour as the blood is driven away and then it will be quickly restored as soon as the pressure is released. Should a toe or finger nail fail to give a satisfactory response it would suggest that one of the lashings securing the splint is too tight and must be eased.

## Unconscious Patients

An unconscious person, no matter what the cause must be protected from asphyxiation. When a patient becomes unconscious the muscles relax including those that control the jaw area. If a patient is lying on his back the relaxed muscles will allow the tongue to fall to the back of the throat and seal off the wind pipe just as it does when we swallow food or drink. With the wind pipe closed asphyxiation must follow. Do the patient a favour and turn him onto his side and with luck he will resume breathing. An added bonus is that if he is sick whilst he is in this position the fluid will drain out of his mouth and not cause him to drown in his own vomit.

Anyone in need of artificial respiration will be in this totally relaxed state but to perform mouth to mouth resuscitation it is necessary to have the patient lying on his back so other steps must be taken to keep the airway unblocked. A pillow, or something of this sort of size, placed under the shoulders will cause the head to fall back and the throat and neck to be arched and stretched. This is usually enough to clear the airway but to make doubly sure take the chin between thumb and finger and lift the jaw slightly. This will also open the patient's mouth.

Place your mouth over the patient's open mouth, at the same time blocking his nostrils with your cheek or by pinching his nose with your disengaged finger and thumb. Blow into the patient's mouth and if all is well you should see his chest rise. Remove your lips from his mouth and the chest will fall expelling air as it does so. Repeat these moves slowly and rhythmically until such time as the patient starts to breath unaided. When natural breathing is re-established turn the patient onto his side into the recovery position.

110

If, when you blow into the patient's mouth, the chest does not rise then clearly air is not reaching his lungs. The most likely cause will be that the jaw is not lifted sufficiently or the pad under the shoulders is too small, thereby not arching the neck enough to clear the airway. Failing this you must look for a foreign body in the mouth that is blocking the wind pipe and remove it.

If you go to the assistance of someone in the water and find that he is not breathing, it is perfectly possible to support the victim so that he is upright in the water with his head falling backwards and give an occasional lungful of air as you pause in your swim for safety. Speedy action is essential when a person ceases to breathe. Without a continuing supply of oxygen the human brain is rapidly damaged beyond any hope of recovery. Three minutes is usually quoted as the maximum period of grace.

It matters not why the patient has stopped breathing, obviously we think in terms of drowning but there can be a number of other reasons for a patient failing to breath, if the patient is not breathing apply mouth to mouth resuscitation at once.

Electrocution will cause the patient's breathing to fail. Make sure the current is shut off before you touch him or you will join the ranks of the unbreathing.

Mouth to mouth resuscitation is as simple as the description given above so even if you have no other form of instruction don't hesitate to use it if the need arises. Voluntary medical organisations often run short courses on resuscitation for non-members and they usually include practice on a dummy. The same courses often include instruction in dealing with cardiac arrest, which is not a subject to be learned from a book.

**Medical Chest**

You will need a medical chest of a kind you never need at home. There you have your doctor at the end of your telephone to come to your rescue. There follows a list of the drugs we were glad to have at some time in the course of our circumnavigation either for ourselves or for a fellow yachtie. It is in addition to a normal first-aid kit.

Some very strong pain killers.

A variety of antibiotics including an antibiotic eye ointment.

Anaesthetic eyedrops or tablets will allow you to remove foreign bodies from an eye that are too firmly embedded or too painful for you to deal with in any other way.

Lomotil was useful to curb severe diarrhoea; there are several other remedies for this that do not require a prescription.

Stemetil suppositories to control a bad case of vomiting are worth carrying.

Spray on burn treatment.

Plastic skin. Cuts and scratches are very reluctant to heal when afloat in the tropics and quickly become infected. Whether this is something inherent in tropical conditions or the effect of repeated contact with sea water which is teeming with microscopic life I don't know, but every passage maker we met complained of this. There are spray cans of plastic skin available and this we found to be most effective in promoting rapid healing and keeping infection at bay. This same material was also useful in treating minor burns.

In addition we carried a stock of sterile dressings and cotton wool, sterile saline and some sterile scalpels, needles with sutures, plastic pots, orange sticks and dissecting forceps. We also had a spray can of local anaesthetic. More about these items shortly.

Doctor David Lewis' book, *Daughters of Three Oceans* has a most informative chapter on what I have called extended first-aid. Although it would not surprise me if the drugs he suggests have been superseded by later development, the article is so pertinent to our situation, he being both a medical man and a long distance sailor. The book is out of print but a Public Library may be able to find a copy for you. It is well worth reading.

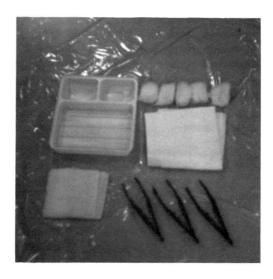

Forceps and large padded bandages with a suitable tray
may be useful for larger wounds.

## Fluid Loss

It takes most people two or three months for their bodies to become accustomed to daily tropical temperatures and it makes sense not to over-exert yourself until you give your body a chance to become acclimatised. By the end of your acclimatisation period your body will have learned to reduce the loss of essential chemicals when you perspire. Even so, extremely high temperatures endured for some time will overcome your defences. I suffered in Saudi Arabia when it was necessary to work hard to repair our boat in temperatures of 110°F and we had been in the tropics for five years by then.

Not only excessive perspiration but extensive burns, prolonged diarrhoea or vomiting can all lead to damaging fluid loss. Obviously the patient should drink lots of fluids to replace the loss of water but this is only part of the treatment that is required. Dehydration will result in the loss of salt and other chemicals that are essential to our well being. Sometimes this loss manifests itself in extremely painful muscular cramps in the hands, feet and legs.

There are powders on the market that are intended to rehydrate patients who have suffered severe diarrhoea, but they will serve to rehydrate a patient no matter what has caused the dehydration. Most contain sodium bicarbonate, potassium chloride, sodium chloride and glucose. They are readily available in most tropical towns. Relief is quite rapid. Within a few minutes of drinking a glass of rehydrating fluid the cramps normally disappear. I would recommend the unflavoured variety ... that is, unless you actually enjoy plastic fruit flavours.

If you don't have this remedy on board, a small quantity of table salt (less than half a teaspoonful) in a glass of water or fruit juice with some sugar will help. Don't overdo the salt, too much and it becomes an emetic.

**Overheating**
The normal body temperature for a healthy human being is 98.4°F or 38°C. Should a patient register a temperature that goes much above 104°F or 40°C, (105°F or 40°.5C is becoming critical) steps must be taken to cool him down. This can be done by repeatedly sponging him down with cold water or by covering the naked patient with a wet sheet which must be rewetted as it dries. It also helps if pads of fabric are soaked in alcohol and applied to the temples and the sides of the neck. In both cases the alcohol pads are placed in areas where large blood vessels come close to the skin. Both the alcohol and the water treatments are intended to utilise the process known as the latent heat of vaporisation. In other words when a liquid vaporises it must take heat from its surroundings to do so, thereby, in this case, cooling the blood stream.

The underlying cause for the rise in temperature could be too long an exposure to the sun or an infection of some sort. In either case the diagnosis is best left to a doctor if you have one within reach. Failing that circumstances could give you a clue. Has the patient been working too hard and too long in the sun? In which case it will probably be enough to continue to work at the temperature lowering and to rehydrate the patient. If over exposure is not the cause then clearly you must suspect that an infection is to blame and an antibiotic is called for as well as the continuation of the cooling processes and rehydration.

Discuss the use of antibiotics with your doctor. There are so many varieties available now, many more suited to certain tasks than others, that expert advice is essential.

**Skin Cancer**
This is a very real danger in the tropics and the fairer your skin the greater the danger. Fortunately it is not usually like its big brother that attacks vital organs and often requires major surgery in an attempt to check its ravages. If skin cancer is not neglected the chances are that the treatment is quick, painless and effective.

By the time we had reached Tonga I had developed a small spot that would not heal on each of my ears. Some months later, in Brisbane, the spots were still unhealed so I presented myself to the Out Patients' Department of The Royal Brisbane Hospital. The treatment was indeed quick, painless and effective. A small spot or sore that is reluctant to heal is the danger signal and your response should be to visit the first hospital that becomes available to you. It should be no surprise that Australian doctors have a lot of expertise and experience with skin cancer. Australia has a large European population and much of its territory is in the tropical zone. Our good fortune is that they are in the right place for us should we need their services.

**Intestinal Worms**
There are still some places where human excrement is used to fertilise growing crops. In these areas it is unwise to eat raw vegetables, lettuce and cabbage in particular, as they can pass on the ova of intestinal worms. No amount of washing or soaking the vegetables in will render them safe to eat. The only safe way to eat vegetables that have been produced in this way is to cook them first. Should you spot that your faeces contain minute worms the treatment is simple and will be available over the counter of a local pharmacy.

You will have to buy the drugs you take with you. If they cannot be bought over the counter you will need a prescription or authority from your doctor. A point to remember is that a prescription is only valid in its country of origin but in many third world countries it is possible to buy drugs without a prescription. There is no such thing as an international prescription.

## Regular Medication

If you are on regular medication you must keep a good stock in hand. Don't forget to include at least three months tablets in your 'going away bag', sometimes vulgarly referred to as your 'panic bag'. In countries where a prescription is needed for your favourite daily drugs, I found doctors in hospitals to be much more sympathetic and very much cheaper than those in private practice.

The alternative is to make an arrangement with your GP to supply a friend or relative with a private prescription and for that good soul to post them to you. Whatever they do they must not label them as drugs but as medicines. I must say that the thought of an international drugs smuggler sending his wares by post labelled 'DRUGS' is an intriguing one, but I assure you it can cause trouble. Added to which much of the international postal system does not approach the high standards of our own service. Far better to buy a stock when and wherever possible.

Check on the shelf life of the drugs you buy. The pharmacist can help you on this score, if his containers do not state a shelf life then it is in excess of three years. Tablets will need to be protected from physical damage by filling any empty space in a bottle with cotton wool. Drug companies cannot dispense drugs to the public no matter how large a quantity you may want.

## Minor Surgery

Minor DIY surgery is not something to be recommended lightly, but if the alternative is waiting twenty or thirty days for skilled assistance it may be forced on you.

We came very close to this situation when we left Darwin for Christmas Island. Just before leaving I had slipped and dealt my shin a savage blow that raised a swelling. In the belief that it would go away we took our departure. We left Darwin in the lightest of winds and drifted our way westward all day. There seemed to be little point in sitting up all night to nurse *Didycoy* a few miles towards our objective so we worked our way into a delightful little bay and dropped the hook. Next morning my leg was worse rather than better and as the day wore on so the swelling got larger. As the size of the swelling increased so did the pain until the following morning it was obvious that we were going nowhere other than back to Darwin.

We motored back and the casualty officer at the base hospital in Darwin opened and drained the swelling giving me instant relief from the pain which had become well nigh unbearable by this time. The wind in this part of the Timor Sea can blow quite strongly to the West and if it had been blowing like that from the time we left we would have been at least 250 miles from Darwin and our return would have been dead up-wind. The thought of cutting into my own leg was not a pleasant one but the pain was such that there would have been no alternative. Long before we got back to Darwin I would have been biting pieces out of *Didycoy*'s toe-rail. I am very glad we were able to get back to professional help but this incident convinced me that it would be wrong to omit any reference to this aspect of medical care.

It seems to me that there are just three situations when it may be necessary to indulge in minor surgery, a wound that cannot be pulled together satisfactorily by any other means than stitching, a swelling of the kind that I have just described or a swelling that is caused by an infection that will not respond to antibiotics and is clearly getting worse.

There are two principles that must be understood before you attempt any form of minor surgery, Antisepsis and Asepsis. Antiseptic treatment is the one with which you are likely to be familiar. It is the way minor cuts and abrasions are treated. This treatment accepts that it is likely that some form of infection has entered the wound with the initial injury. The usual treatment is to bathe or swab the wound with an antiseptic solution and then to cover it with sterile gauze on which may have been spread an antiseptic cream. Note that the aim is to kill any bacteria that may have invaded the site.

Aseptic treatment on the other hand aims at not introducing infection into the wound (the wound being the incision or the punctures made by the needle). This aim is achieved by using instruments and dressings that have been sterilised and have since remained uncontaminated. It will be your responsibility to see that this gear remains in that state while you use it. Ideally you will need an assistant. This may not be possible and you will have to do both jobs. It will make for ease of explanation if we assume that there are to be two of you.

The area which is to be operated on must first be cleaned and then swabbed with an antiseptic solution. When the antiseptic has dried the area must be anaesthetised using a spray on local anaesthetic. There will be instructions on the can. With everything assembled you must scrub your hands for several minutes with soap and running water.

Once this is completed you must not touch anything that is not sterile. Your assistant should open the packets you need without touching anything that is inside the packets. You must extract the contents without touching the outside of the packets. The first thing you will need will be a large sterile paper towel which must be placed on the table where you are to work. The contents of successive packets must be placed on this sterile paper towel. This is the way you must proceed, as if there were two worlds, the sterile and the unsterile and never the twain shall meet.

The method to be used for inserting stitches and opening an uninfected swelling should be self-evident. You are unlikely to meet up with the sort of swelling I suffered. I happen to be a special case. I am on a daily dose of warfarin which reduces the clotting power of the blood. When I damaged my shin the blood that was leaking within my damaged tissue failed to clot and so for about 72 hours the swelling continued to grow unchecked.

An infected swelling should respond to antibiotic treatment so you are likely to be cheated again. In case things go wrong and you are too far from medical help the following points may be of value.

Do not assume that because the swelling that you are about to open is infected you need not observe the rules of asepsis. Paint the surface with a good antiseptic and anaesthetise.

The centre of the swelling is the likely seat of the pus which you aim to drain away and this will be surrounded by a hard wall, the prophylactic wall. On no account should this wall be damaged in any way. It is nature's way of containing the infection and if it is breached the infection is free to invade the rest of the body with possibly serious results. The object of the exercise is to penetrate the skin above the pus to release it and to allow the resulting wound to be drained of pus and flushed with a suitable antiseptic.

Don't rush into surgery of this kind until antibiotics have been given a chance to work which they almost certainly will. The signs of failure will be a reddening of the area, the same area will be hot to the touch and a red line will begin to move up the limb from the site of infection.

And finally, if you have a ham radio on board you will be able to call for advice and maybe even assistance. It is usual these days for Ships Masters to have undertaken a course of study in hospital so your cry for help could fall on informed ears.

Surf on a reef between Prison and Direction Islands on the Cocos Keeling group.

## Chapter 15
# Fishing

Even if you are a better fisherman than I am, and most people are, you will not live off the fish of the oceans, but what you do catch will be a very welcome addition to your larder.

The oceanic fish are great muscle machines and you must have gear strong enough to keep them if they are hooked. In this kind of fishing one is not indulging in a contest to see who can catch the biggest fish on the lightest line, one is playing for keeps. We started with a monofilament that had a breaking strain of 90lbs. When we crossed the Atlantic we lost so many fish that I upgraded the line to 140lbs as soon as we reached Barbados. We increased the strength of the line twice before we kept what we hooked. By this time we were using a line with a breaking strain of 280lbs!

We used a lure that measured no more than $3\frac{1}{2}$ in in length in the mistaken belief that we would only attract modest sized fish with such a small lure. We were so wrong. The biggest we caught on that bait measured nearly five feet in length and a three foot long fish was not at all unusual. I know it sounds like a fisherman's story, but it is quite true that we had a small number of hooks straightened by fish that escaped from our 280lb breaking-strain line. What size they were I don't know but I was not sorry to lose them. The thought of sharing the cockpit with an angry creature that could do that was to say the least, daunting.

We used double hooks about $2\frac{1}{4}$ in long, the metal was nearly one eighth of an inch thick on both shanks. Our lures were small bundles of coloured strips of bunting no more than $3\frac{1}{2}$ in long. Again and again we were surprised at the smallness of the mouths of the fish we caught. The hook was attached to a short length of wire of similar breaking-strain to our monofilament. This was intended to prevent the fish biting through the monofilament. The wire in turn, was connected to at least 150 feet of monofilament so that we could get the hook well away from the boat.

At the inboard end the line was attached to a large wooden reel. (Figure 32) The vertical axle was mounted on the side of the cockpit coaming so that the reel was free to rotate in the horizontal plane. One end of some heavy shock cord was arranged as shown in the diagram and the other end was made fast to a strong point in the cockpit.

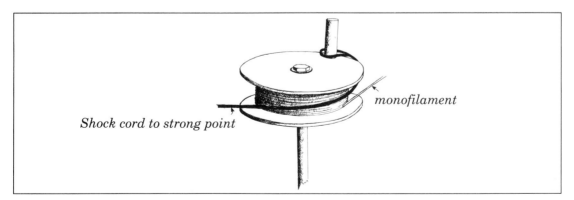

*Shock cord to strong point*    *monofilament*

Fig 32. The all important reel on the ship's end of the line.

If a fish struck it would attempt to race away as soon as it felt the hooks in its mouth. The shock cord would reduce the impact of the initial strike. This would cause the reel to rattle and bang in its slack fittings so sounding the alarm. Then it was a matter of reeling in. If the fish was big and powerful this was a slow business as it was only possible to reel in when the line went slack as the fish ranged about in an effort to break free. With the fish alongside the next problem was to get it aboard and a big one could be difficult, half a hundredweight of muscle is quite a handful and some of those teeth had to be seen to be believed.

The real problem moment, the moment when most fish are lost, is when the fish is lifted high enough out of the water for its gills to be exposed to the air. In human terms this must be akin to drowning. The fish always makes a wild fight to get free at this point. If you manage to get the poor creature into the cockpit it does nothing to reassure the fish. Twenty to fifty pounds of angry muscle armed with a mouthful of very sharp teeth and a number of very sharp spines on tail and dorsal fin in close proximity to your bare feet is not the best companion with which to share a small cockpit.

The usual drill, which can take some time, is to use a winch handle to beat the poor fish to death. Fortunately there is a more humane and elegant answer. With the fish held vertically alongside so that its jaws were just out of the water it would lie quietly. My wife would take a small aspirin bottle full of rum, gin or scotch which we kept handy in the cockpit and empty it into the fish's mouth. This would produce instant unconsciousness on the part of the fish which could then be hauled onto the side deck and despatched by cutting off it's head. I know it sounds like the sort of story that old salts tell young seamen .... the tale of the golden rivet and all that .... but it happens to be true; try it and if it doesn't work you can have your fingers back.

A stout pair of leather gloves makes it easier to hold a slippery fish when you are dragging it aboard and they will protect your hands against puncture wounds from the very sharp spines many fish have as part of their dorsal fins and sometimes their tails too. If you are going to use my tot of rum tip you will not need a gaff as an aid to landing your catch.

A gaff hook helps get dinner aboard.

A fish that is large enough makes it possible to run a sharp knife down either side of the backbone and remove a large steak without having to indulge in gutting your prize which is the least pleasant part of the whole process. It is when you catch a good sized fish that you realise the benefit to be gained from having a small freezer aboard. There is a limit to how much fish two people can eat before it begins to go off in the tropical heat.

For the last two or three hundred miles of our Barrier Reef sailing we were in the company of an English couple in a boat named *Sea Loone* and they were wonderful fishermen. If, by the time we were anchored each evening, they had not caught enough fish for us and the Dutch boat *Klepple* who was also with us, they would often put a net across a break in the fringing reef. They would choose a stretch of reef that was close to and parallel with the shore with shallow water behind it. Pan-sized fish that were hunting the smaller fish behind the reef would enter on a rising tide and leave before the falling tide left them high and dry. With the net in place some time after high water the fish our friends were after would attempt to leave and become entangled in the net that was stretched across the opening.

Their other ploy was to trail a light line and a small hook and lure some distance behind their outboard-driven dinghy. How they chose their fishing grounds for this operation I never did discover. I can only report that when I tried it I caught nothing. They were always successful.

Just about the right size.

# Chapter 16
# Fresh Water

We are so accustomed to unlimited supplies of safe fresh water at home and on the nearby continental coasts that it comes as a surprise to discover that it can be several thousand miles to the next hose pipe of sweet water. From Cristobal Colon to Australia, just about the full width of the South Pacific, about 9,000 miles, we were able to fill our tank from a hose on three occasions, Cristobal Colon, Tahiti and Tonga'tapu. It is true that there were a number of places where we could pick up fresh water in jerricans and transport it back aboard by dinghy but in many places there was no public supply for the passing yachtie, or, as in the Galapagos Islands, the water was unfit to drink unless you had been weaned on it.

We have a sixty gallon tank onboard *Didycoy* and carried another twenty gallons in plastic bottles as a reserve, particularly with the thought in mind that the tank could spring a leak. Our 60 gallon tank was more than adequate for the two of us, which is not to say that we crossed the Pacific with only two refills. At sea we managed on a gallon a day each without hardship. If we were restricted to this, showers were not possible and we had to resort to the bird bath and face flannel routine. It was unusual to be restricted in this way for there is an abundant free supply of pure sweet water from the skies above. If you have made suitable provision for collecting this free delivery you can usually fill your tanks at least once on a thirty-day passage.

When it rains in the tropics it is something special. Visibility is reduced, often to two or three boat lengths by the sheer volume of water that is falling. I remember that in one unexceptional rain storm we were collecting water at a rate of one gallon every forty seconds from an awning that measured a mere nine feet by seven.

It is often recommended that when sailing in heavy rain the boom should be lifted a little with the topping lift and the resultant waterfall at the goose neck collected in buckets. Fine, so far as it goes, but why go out in the rain and cart buckets of water back to your cockpit or wherever? The rain is usually ice cold, after all it has fallen from great heights and the individual drops are often heavy enough be quite painful when they strike bare flesh.

I screwed lengths of 2in x 1in timber to the edges of our main coach roof making a 2in high wall all round it. From the two forward corners two short lengths of electrician's plastic conduit descended and joined a horizontal run of the same piping that conducted the rain water to our tank. The first water that fell was always allowed to run over the side until it ran clean and free of salt, then it was diverted into the tank. I used ³/₄ in diameter conduit. The bends and tee junctions that are designed to go with it are a simple push fit and form a joint that is sufficiently water tight for this purpose.

If you adopt this system of water collection and have a canvas or sailcloth sun awning for protection when you are at anchor it will keep the rain from your collection area. Three or four plastic skin fittings incorporated in your awning at strategic points will solve the problem.

The only place we were unable to collect rain water was the Red Sea, where it seldom rains. It created no problem because the Arabs were so very good to us on this score.

The 'dam' around the edge of the coach roof.

If you feel that converting your coach roof is too extreme a solution for your boat, I have seen a yacht with a very large collar that could be strapped around the mast just below the goose neck. This fitting terminated in a length of pipe from which water bottles could be filled if the boom was well topped up. It was not as efficient as the coachroof wall, it had to be rigged whenever it was needed and did not serve when the sails were stowed. This system could be improved by running a line of piping from the mast to the tank filling point. At the mast, the pipe from the funnel could be plugged into the pipe whenever it was in use. However, whatever solution you prefer you must make provision for the collection of water both when you are at anchor and when you are under way.

A charcoal water filter is well worth installing. In many places the water supply is treated chemically and has a strong chemical flavour. A charcoal filter will do much to remove this unpleasant taste. These filters are not everlasting so you must take replacements with you. It is likely that after leaving Europe you will not find any more filters until you reach Australia or New Zealand.

In common with others we found that we could expect a charcoal filter to last for two years, whether it was still functioning as well as a new one would I have no idea, but as we were only depending on it to remove the taste of chemicals we were happy to use it for as long as it did just that. On one occasion I ran tapwater the wrong way through a time expired filter that was clogged. This cleared a lot of gubbins out, mostly algae, and the filter was returned to service.

If you do fit a filter and you have an electric water pump, position the filter between the tank and the pump. That way you wont be plagued with pump problems caused by small particles of detritus collecting in it, the filter will remove them before they reach it. If you intend to fit an electric water pump keep it in mind that there two kinds, those that burn out if they are run dry and those that do not. The latter type are usually a little more expensive but you normally only need to buy them once!

We found, with very few exceptions indeed, that the water supply wherever we were was normally quite pure. If you have reason to doubt the purity of the water supply it is best left alone. Chlorine and other chemicals in the proportions usually used in drinking water will not kill the cyst of the dysenteric amoebae, nor will it touch some other equally harmful organisms. Boiling water vigorously for at least five minutes is sometimes recommended as an answer to the problem but I have learned that some pathogens can resist even this treatment. The answer must be if the source of water is suspect don't touch it.

Reverse osmosis is a form of filtration that has come into prominence in recent years and it is efficient enough to be used for desalination for large areas of the Middle East and elsewhere. Small units are on sale for use in yachts and we met a number of American boats that relied exclusively on reverse osmosis to convert sea water to drinking water. I have talked to professionals in this field and they have said that so long as the equipment is in absolutely perfect condition the water produced is likely to be free from all forms of contamination. Note the use of the word 'likely', they were not willing to give a categorical assurance on this point.

When failure occurs it could be as simple as an 'O' ring that does not quite do its job and allows maybe a minute quantity of contaminated water to by-pass the filter. This could be enough to nullify the successful filtration of the rest of the water. Filtration plants will constantly test their product for contamination, something that we are unable to do.

Rubber or neoprene water tanks are on sale for installation in boats. They are not for passage makers. If you install one you will be lucky to get across the Atlantic without chafing a hole in it.

A view from the dog-house of the rain replenishing our fresh water supply from the coach roof dam.

Chapter 17
# Keeping Cool

Ventilation and shade are the keys to a cool boat in the tropics, you really cannot have too much of either. In harbour or at anchor an awning and a wind, sail will go far towards making life comfortable for you. The more of the decks and coach roof that can be put into shade by the awning the better. We found that within ten minutes of erecting the awning the temperature in the cabin dropped noticeably. This will come as no surprise when you realise that if we were stationary without the awning rigged the deck was often too hot to walk on barefooted. I am sure that the breeze passing over the boat but under the awning helped the cooling process.

## Awning Material

It is a problem to know just what material to use for the awning, they all have their drawbacks.

Canvas has been the favoured material in the past but it has two disadvantages. It must be totally dry when it is stowed or it will rot. To be strong enough to do the job canvas needs to be of a fairly heavy grade and this makes for a bulky bundle when it is rolled up. Its great advantage is that it makes very little noise in a wind.

Sailcloth can create quite a noise in a strong breeze and the trade winds blow at a good 20 knots for much of the time. Man-made sailcloths have the advantage of not rotting if they are stowed in a damp condition but they do sometimes develop fungal spots that are disfiguring.

In Australia we found what must be the nearest thing to the ideal material. It is called shade cloth and is an open weave plastic material that does not rot or stain even if stowed when wet. It comes in various grades measured by the percentage of sun it keeps out, we used the 90 per cent variety. Because this material lets the wind blow through it does not rattle in the wind in the way that sailcloth does, nor does it need to be taken down if the wind begins to pipe up. On numerous occasions we have sat under our awning watching the rest of the anchorage struggling to get their awnings down before the sudden increase in the wind damaged them. Shade cloth is often available in garden centres.

The edges of the shade cloth need to be reinforced to allow the awning to accept the large eyelets that are necessary to hold the tying down lanyards and we found two inch wide deck chair webbing to be quite suitable for this purpose.

## Wind Sails

We have seen a wide variety of wind sails in use and have tried at least three patterns. Anything is better than nothing but some patterns are definitely superior to others. The form we favour is about two metres tall and a metre wide. In use it is cruciform in horizontal section held open in this shape by a cross piece of two battens at the top. A lanyard from each bottom corner made fast to four lacing eyes screwed to the deckhead just inside each corner of the fore-hatch, secures the lower end.

Our wind sail was made of very lightweight spinnaker sailcloth. Two lengths of this material, 2¼ metres long by 1 metre wide, must be laid precisely one upon the other and sewn together along the centre line for two metres of their length. To give some extra strength to this seam, machine in one or more parallel lines of stitching. The remaining material will be used to make sleeves to contain the two cross battens, do not cut it off. The two battens need to be about 2in wide by $\frac{1}{2}$ in thick and when in position they must extend beyond the fabric by about two inches at each end. One will remain in the sail and the other must be housed so that it can be removed to allow you to roll the fabric around the battens when you wish to stow it.

Remove the sharp edges of the wood with a plane or glass paper and taper the removable batten slightly to make it easy to house it in its sleeve. Drill a hole about an inch in from each end of the battens to take an end of the four lines that form the bridle with which the wind sail will be suspended. Where the battens cross you will need to drill two small holes to enable you to bolt them together when the sail is in use. Bearing in mind that one batten needs to be just above the other, construct the two batten sleeves with the excess material. Reinforce each lower corner with at least two layers of material and insert an eyelet into each one. Splice a lanyard into each eyelet. It now remains to 'roof in' each of the four triangles at the top of the wind sail with sailcloth.

When the sail is raised it will probably be necessary to use a lanyard to pull the top of the sail towards the forestay to ensure that the sail stands upright. No matter how variable the wind may be there will always be one face of this wind sail waiting to intercept it. Fabric a metre wide usually allows enough material in each arm to permit the two parts to drop back when full of wind so that the centre of the sail just touches the far side of the hatch. If your hatch is small it might pay to taper the sail from top to bottom in the hope that it will increase the volume of the descending air.

The only time this wind sail will fail to push air through your boat is when the wind is from astern. If the wind blows in through the open main cabin/cockpit hatch it will meet the air from the wind sail and both will come to a stop. When this happens the only answer is drop the sail and let the air from aft take over.

For windless occasions we installed a 12 volt fan, purchased from a caravan accessory shop. Incidently, these shops are well worth a visit if you are fitting out. The fan was a beautiful bit of luxury but alas it only lasted about two years before it expired.

## General Hints on Keeping Cool

A white hull will reflect some of the heat rather than absorb it as would darker colours. Indeed white is almost a mandatory colour for a wooden boat in hot climates if you wish to avoid the seams opening up. It would be good to have white decks too but white decks in tropical sunlight can be hard on the eyes and so it is more usual to have light coloured decks.

Cotton is the ideal material for clothing. Polyester and nylon should be avoided, perspiration soon makes garments made of these materials wet and uncomfortable to wear. Mixed fibres of cotton and polyester are fine so long as cotton predominates. Clothing that is light in colour and loose fitting will also help to keep you cool.

The human head is plentifully supplied with blood vessels, many near the surface. This, we are told, leads to considerable heat loss in cold weather and we are advised to wear a woolly hat to reduce the loss. It seems reasonable therefore to reverse this line of thought and say that it is very likely that for the same reason we can absorb a great deal of heat through the top of our head. If you are thin on top, or worse still, like me thin on top and very fair .... eggshell blonde is the term I think .... some kind of hat is a necessity. The French Foreign Legion had the right idea with the 'kepi' their troops wore, it helps to protect the back of the neck.

When you feel that you need to be cooled down, a cold shower is the perfect answer, especially if you refrain from drying yourself. The evaporation of the shower water will go a long way towards cooling you. Soaking your hair in water and your hat too works wonders. Some sort of shower on board is an absolute boon. An elaborate pressurised water system is not the only way to deal with the problem, there are simpler and cheaper alternatives.

Caravan supplies shops sell a minute 12 volt, submersible electric pump that can be dropped into a plastic jerrycan of sun-warmed water to operate a shower head. Garden centres sell a weed killer/fertiliser spray bottle that makes a good shower. All that is necessary is to operate the hand pump to compress the air in the bottle and open the valve on the spray head. Purpose made showers of this kind are appearing on the market now and they are very good. If you can secure the canister in the sun your water is warmed free of charge. A self-draining cockpit makes a simple shower room. There is also on sale a strong, black plastic bag with a shower head attached. Filled with sun-warmed water and hung from the boom it makes an excellent shower.

The wind sail in position above the fore-hatch. Its four section construction ensures that the wind from any direction will be redirected into the boat.

# Chapter 18
# Insect Pests

In the insect world you will have three principle enemies, Mosquitos, Sand flies (No-Nos or No-Seeums) and Cockroaches. The domestic fly can be a nuisance in some places but at least it does not bite and it is easy to deal with, a plastic whacker being our favoured weapon. It may help you to know that the common house fly takes off backwards so aim astern rather than ahead when you use a whacker.

There are two ways to deal with flying pests, the old and the new. We will deal with the old first. Usually you will be anchored far enough offshore to be out of reach of most of the insect life but when you are close enough to the shore to be bothered then you will need screens for all the cabin openings. Items like dorades, mushroom ventilators and sliding windows can have a piece of plastic mosquito netting installed and then be forgotten.

Hatches and entrances must have their own made to measure screens. They must be easy to rig or they will be omitted and you will suffer. Port holes are a problem and the best solution we could devise was to attach a disc of stiff plastic screen to the port hole surround with three generous blobs of Blu Tac.

Caravan shops sell Mosquito Coils. These are coils of I know not what, which, when set up and ignited will smoulder for about six hours and seem to keep the beasts at bay. The trouble is that if they are used frequently they cause the boat to smell like a charcoal burner's hut. If a mosquito or a sand-fly comes aboard it is likely to hide in some dark corner until it is dark when it will come out to torment you. The only answer is a comprehensive spray of insecticide and evacuate the cabin until it settles.

If you are bitten by a mosquito or a no-no, an antihistamine cream will often go far to stop the intolerable itch.

All this was routine when we met with the pesky critters until one day well on into our circumnavigation we were introduced to the ultimate deterrent. Overnight, screens, aerosol sprays, plastic whackers and size twelve deck shoes became a thing of the past.

There is on the market a small gadget that when plugged into a 240 volt circuit will gently heat a small pyrethrum tablet. The tablet will give off just enough vapour for eight hours to keep the average cabin free from flying pests, doors and windows open. We are assured that pyrethrum is harmless to human beings. Similar gadgets that will work from 12 volts DC are appearing in caravan shops. We found that the tablets were available in many of the tropical countries we visited. The twelve volt model was not available when we discovered this little machine and so we had to work out our own design and I set it out here in case you cannot find the purpose-made thing.

The heart of the heater is a pair of resistors that create the correct heat when a twelve volt current is applied to them.

The resistors are marked thus:

$$\vdash \text{238-4}$$
$$\vdash \text{16R 5\%}$$

They are used in TV repairs and it is to a TV repair shop you must go to find them. They are not expensive. When you get them the wires will not look quite as they do in the sketch, they will extend for about one and a half inches from each end. One wire on each

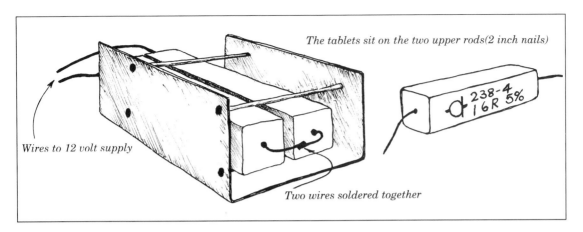

*The tablets sit on the two upper rods(2 inch nails)*

*Wires to 12 volt supply*

238-4
16R 5%

*Two wires soldered together*

Fig 33. The DIY electric insect zapper.

resistor must be reduced to about half an inch in length and these two short wires must be soldered together. The remaining longer wires must be connected to the positive and negative wires of your ship's supply of 12 volts DC.

My own model is a bodger's delight and I have promised myself that one day I will make a more elegant job of it but in the meantime it works. The body is a two inch long piece of aluminium extrusion found on a boat yard rubbish heap. Two nails span the bottom of the box so that the resistors can rest on them to allow air to flow past the resistors. Two more nails span the top of the extrusion to make a grid on which the tablet sits to be warmed by the heat from the resistors.

## Cockroaches

If you stay in the tropics for any length of time you are almost certain to play host to your share of cockroaches. Their appearance on board seems to provoke two emotions in most people, distaste that amounts almost to horror and a sense of guilt that somehow you have failed to keep your boat as clean as you should. The cleanliness of your boat is in no way to blame for the appearance of these little horrors.

They will come aboard either as eggs or as live specimens in the folds of packages, cartons and boxes. There then is your first line of defence, none of these items must come aboard once you reach subtropical latitudes. A favourite hiding place for young roaches .... pin head size or smaller .... is between adjacent bananas where the two stalks meet. You must wash all fruit and veg the moment it comes aboard. Whilst it is true that the adult is about three quarters of an inch long the young are minute and can easily escape detection. If you are close to the shore the adults will glide aboard if the wind is in their favour and outflank all your precautions, we've watched it happen. We stayed clear of cockroaches for three years and it was not until we reached that cleanest of all countries, Australia, that we became infested with the brutes. It took eighteen months to clear our boat of them, largely because of our ignorance.

We tried everything. We left little heaps of boric acid in remote places, scattered poison bait traps hither and yon, regularly sprayed surfaces with residual insecticide and left beer cans with half an inch of beer in the bottom. They all worked and we remained infested. When we sprayed a roach it would disappear into the bilge and die. Honour was satisfied, we could always pump the dead ones out with the bilge water. We had been told that when an adult cockroach is killed it lays a sac of eggs as its last act. We laughed this off as an old wive's tale. When we actually saw it happen we stopped laughing and realised why we were not winning the battle.

I cut a hole in the top of a plastic cap that comes with an aerosol can and slapped the cap over each adult roach we caught. A burst of insecticide was sprayed into the cap through the hole and the cockroach died where we could pick up it and the egg sac and put them over the side. This also had the effect of limiting the amount of insecticide that we inhaled. Then we began to feel that we were winning. About two in every three adults we caught laid an egg sac as it died so it was no wonder that we had been losing.

I did some research in a public library and found that each sac could contain as many as forty eight eggs! On the good news side it takes about ten months before each cockroach is mature enough to start laying eggs so once you have decimated the adults you have ample time to deal with the youngsters. It does take bloody minded persistence to get rid of the pests. They are creatures of the dark principally and so each evening I would make at least three forays around the boat with a torch, the plastic cap and a can of insecticide; they became known as 'Captain's Rounds'.

We began to believe that we were winning when we were only finding immature specimens and we were so wrong. Betty found a store that sold 'cockroach houses' and she brought half a dozen back with her from shopping one day. They are really oversized adhesive flypapers on a stiff card base. We laid those six traps out that evening and by the next morning they had caught 165 cockroaches.

Many of those we caught by this means that night were adults with egg sacs. I recall that one egg sac had matured and produced a mass of young, each one much smaller than a pinhead. These youngsters were stuck to the flypaper in a wedge that was heading towards the edge of the trap. These minute specimens were not included in our body count as they were too small to count but at a guess there must have been forty or more of them which would bring our catch in one night to well over 200 and this did not include the large number of unhatched eggs.

We had one last weapon to try, the bomb. This is a can of Gammexane which is distributed as a lethal smoke when the wick is ignited. Before this is done all unsealed food must be removed from the boat, floorboards lifted, locker doors opened, sail bags emptied and so on. Then all openings must be sealed, your escape route planned and then the bomb may be lit. You must then stay around to fend off people who will want to smash open your hatch so that they can get to the source of what they think is a fire. Knowing what we know now we would not mess about with sticking plaster remedies, we would go straight to the bomb and this is what I would urge you to do.

Once you have seen forty or fifty of the minute brutes stream out of one, repeat ONE egg sac you realise that anything short of the bomb doesn't stand a chance. Of the 165 roaches we caught that night there were at least 50 of them that were carrying and ejecting egg sacs.

## Chapter 19
# Domestic Arrangements

**Cookers**

Prior to our departure, the collective wisdom on cookers dictated a kerosene model for long distance sailing. It was generally agreed that gas was often difficult to find and when you could find it was likely that there would be difficulties finding connectors that would marry your cylinders to the gas supply. This is no longer the case. For some years, at least as long ago as when we bought our kerosene cooker, gas has been more readily available than kerosene. People of the third world countries are accustomed to improvising and we encountered no difficulties with odd ball connectors.

World-wide the quality of kerosene, when it is available, varies enormously. The poorer quality stuff carbons up the needle valves on the burners and causes trouble, expensive trouble. Our cooker had three burners and an oven and we could easily get through two burners in six months. Spares could be difficult to obtain and were usually expensive when we could find them.

I seemed to spend more time cosseting that stove than any other single item on the boat. The methylated spirits needed to preheat the burners each time they were lit was always expensive and often unobtainable. On occasions we were forced to use precious fluids such as vodka and gin. We found that these liquids were unwilling to ignite without the addition of a small quantity of methylated spirits. Their other characteristic was that once they were burning they would usually continue to burn for a very long time. We have on occasions boiled enough water for two mugs of coffee on one preheating shot of the mixture!

We suffered these problems until we reached Australia and then bought ourselves a gas cooker and have never regretted changing. I am well aware of the dangers inherent in the use of gas aboard a yacht but oddly enough it is kerosene that came very close to ending *Didycoy*'s career by fire. The gas at present available to us is heavier than air and this means that any escape from your gas system within the boat will sink to the bilge and stay there. When there is a sufficient concentration in the bilge the first spark to reach it will blow you sky high. Moral, don't let the gas escape but if it does then ventilate the bilge very thoroughly or you will find yourself acting as starboard look-out on the wettest cloud St. Peter can find for you.

The ideal set-up would be to have the gas bottle secured somewhere on the upper deck with the piping running outside the accommodation to the galley with a cut off valve where the pipe enters the accommodation. With this layout the cut off valve will be used, whereas if the valve is inconveniently placed it could well be ignored. It is vital that your installation should conform to the latest set of recommendations and be well maintained. Understand the dangers and develop a routine that respects them.

The final two or three feet of the gas delivery hose must be flexible to allow the stove to move in its gimbals and this is where chafe is most likely to occur. Inspect the pipe with chafe in mind at regular intervals, remember that you will be sailing for very much longer periods of time than perhaps you have ever before. Carry two or more spare lengths of flexible pipe and do what you can to protect both the pipe in use and the spares against rust. If the official recommendations allow it, use plastic tube for the delivery hose rather than the kind of tubing that has a spiral binding of metal strip. This metal strip is galvanised mild steel and it starts to rust after only a short time at sea.

The cooker must have a lock or restraint of some kind to stop it swinging when it is not in use. This will go some way to reduce the chafe to which the flexible section of the delivery pipe is exposed but it will not eliminate it completely.

Matches are better left ashore if at all possible. They must be kept dry, which is not easy; they do offer a threat of fire in as much as a discarded match may still be burning when you drop it into the rubbish bin and in many places the matches on sale behave like fireworks when they are struck. Gas, meths and preheated kerosene will all ignite with a spark but oil lamps will not. The only answer here is a refillable gas lighter and a small can of gas with which to refill it.

To be without cooking facilities is a major disaster so you must carry an alternative system. Being born pessimists we had two, a small folding Primus stove and a single burner Gaz picnic stove. The refills for our Gaz stove were on sale in many of the places we visited so their replacement presented no problem.

Our emergency cooking facilities served a second purpose. When it was unbearably hot we would use them in the cockpit to avoid overheating the cabin. Not only did we find our spare cookers a God send on a number of occasions but so did several other folk who had cooker problems and had not thought to have a back-up system.

If you use the small Gaz cylinders into which the cooker or lamp fitting must be screwed, do change the cylinders on deck. Apparently empty cylinders often have a residue of gas in them which will escape when the cylinder is detached from the fitting. When fitting a new cylinder screw it home without hesitation. Don't be surprised if you hear the hiss of escaping gas, continue screwing the cylinder home and it will stop. I mention this because a friend of mine who was unfamiliar with these small gas canisters heard the slight hiss of escaping gas as the top of the cylinder was punctured, thought that he had done something wrong and unscrewed the can. The resulting fire ball that flew around the cabin put three of them into hospital.

## Interior Lighting

I suppose that electric lighting is the most agreeable form of illumination and if you have sufficient battery and generating ability I am sure that it is the least bothersome way to go. Gimbaled oil lamps are fine for eating and socialising by but the do not give enough light to allow you to read. The amount of heat they contribute is not enough to be a nuisance.

As a general rule the wider the wick the whiter will be the light. The wick must be trimmed to a slightly convex shape to burn well. After a period of use a hard layer of carbon forms on top of the wick and interferes with the quality of the light. It is a simple matter to rub this off and restore the light to its former level.

Don't let your oil lamps run dry as this burns the wick and then it must be retrimmed before it will burn properly. Do carry some spare lamp glasses with you. Sizes vary all over the place and you may well not be able to find your precise size anywhere.

Gas lamps and kerosene pressure lamps give a very much better light to read by but they both add very considerably to the temperature of the cabin. If you are to use kerosene for your lamps you must give some thought to filling them when under way; a funnel is not enough. Chasing a small oil lamp around a heaving cockpit with a two gallon can is fun for the onlookers but frustrating for the main actor.

I soldered a tap to the base of a one gallon oil can, the tap was the type used for a boat's fuel tank. A length of plastic tubing was attached to the tap and this was long enough, when stowed, to reach above the top of the can; by this means any leakage from the can was contained within the pipe. The can was strapped to a bulkhead in the forepeak with a deep fiddled shelf below to take the lamps when they were being filled. I know that tin plate cans are becoming a thing of the past but it maybe that the above description will help you think your way around the problem. The longer we sailed the more it was driven home to us that good solutions to small domestic details of this kind made life so much pleasanter.

My wife had a Tilly kerosene smoothing iron and we have been very pleased with it. It was not often used but whenever we were invited to meet people ashore or to eat out on some special occasion it has been nice not to be the only people present who were dressed in unironed clothing. I suppose a small old fashioned solid iron that could be heated on the cooker would serve perfectly well and would be very much cheaper.

## Cold Storage

We set out without a fridge/freezer and finally bought one after three years. It could be said that this proves that you can survive without a fridge/freezer but we wouldn't do it again.

Cold drinks are nice but to my mind they are simply a spin off. The real reason for having a fridge/freezer in a boat is to ease the cook's job by providing fresh food when you are on passage. Without a freezer we found that after three days into a passage we were opening cans and this often meant opening cans for the next two or three or more weeks. Without refrigeration we did not dare keep fish for more than two or three days in tropical temperatures, so a lot of otherwise useful food was thrown away. A four foot long Dorado is a lot of fish for two people to eat in two or three days!

If economics dictate that it must be a choice between a freezer or a fridge then it *must* be a freezer. The freezer need not be particularly large, I have seen many built into boats that measured about 12 x 6 x 15 inches and they were giving good service, after all their task is usually only to keep enough meat or fish fresh to satisfy two people. It could be that there are several arguments in favour of limiting the size of a freezer or a fridge that is to be installed in a boat: ease of maintaining a low temperature and lower cost are two that spring to mind.

I have heard it argued that the bilge is cool enough to keep food fresh. In home waters maybe. In tropical waters, no way! Surface sea temperatures in the tropics are commonly 80°F and even reach 90°F at times and the temperature in your bilge will closely reflect those figures. What chance does meat or fish have of staying edible in those temperatures?

If this book is to serve any real purpose at all it should be to help you avoid the mistakes we made and the biggest mistake we made was to assume that a tiny freezer was an unnecessary luxury. There are a number of different types of fridge/freezer units available and it would be nice to be able to say get this one or that, but advice of that kind would be out of date before it was published.

There are two systems that I have seen working at close quarters. One is the so called Eutectic system. This comprises of a well insulated top loading box in which is fitted a stainless steel tank filled with a special fluid. Connected to the engine is a compressor that serves the fridge. When the engine is running the compressor acts on the tank of fluid and causes it to lower its temperature quite quickly. If the engine is run for half an hour, morning and evening, it is sufficient to keep a small fridge and freezer at suitable temperatures in an ambient temperature of about 33° Celsius or 91° Fahrenheit.

My only first-hand experience was impressive. We had gone aboard a boat for a day's sail with Australian friends. The boat had not been used for three or four weeks. The freezer was empty and the lid was not in place. Within ten minutes of starting the engine the interior of the lidless box was coated with ice; remember this was in tropical temperatures.

The other unit I saw in action was again in tropical temperatures. This was a fresh water cooled unit developed in Australia. The cooling water, plus a little bleach, was contained in a screw capped stainless steel tank, the bleach was to inhibit the growth of algae. Copper pipes brought the coolant to the tank and through the water so that it could be cooled. The coolant was pumped around for about eight minutes and then the pump shut off to allow the water that had been heated by the coolant to cool to the ambient temperature, just as a cup of coffee will cool when it is left standing. As soon as the tank of fresh water had cooled sufficiently the pump switched on again if further cooling of the freezer was needed. All this was easily powered by one solar panel. Those yachties we met who were using it reported very favourably on it. Unlike the Eutectic system, which is excellent but requires the engine to be run for thirty minutes once or twice a day, this is literally a switch on and forget system.

Air cooled fridges are inefficient in the tropics for obvious reasons and salt water cooled fridges suffer from corrosion from the salt water that is pumped through them as a coolant. Kerosene and gas fridges are dangerous when the vessel is underway.

To sum up. Top loading, small, over insulated and must be able to freeze rather than just refrigerate are all essentials. This is a field where rapid improvements are being made so check around and see what is available. And finally, at the risk of being repetitious I regretted not building a freezer in before we left. Next time, if the choice is to be between a GPS or a freezer there will be no contest, the freezer will win hands down!

## The Galley

We tend to think of the foredeck as the area where we are exposed to danger and it's true enough, it has its problems. The galley has its dangers too. The dangers are twofold, a pan of boiling liquid that comes off the stove in response to a lurch of the boat and the difficulty of working and maintaining your balance in a seaway. A sudden roll can throw the cook across the cabin and cause serious injury. Both dangers can be greatly reduced by proper design in the galley area.

The cooker must be gimballed with the axis of the gimbals set fore and aft, it is the rolling that needs to be countered. A deep fiddle and pot clamps are also essential.

A PVC apron will go a long way to protect the cook from severe scalds if a pot is thrown off the cooker but the temperatures are often so high as to make this impractical. I recall my wife's anguished antics when, one day, wearing her smallest bikini, she removed the lid of a saucepan of popcorn she was preparing to be met with a sudden shower of exploding corn.

There must be a bar across the galley in front of the cooker to prevent the cook being thrown onto the stove and to give the poor soul something to cling to in moments of crisis.

I would rate the galley layout in Figure 34 the safest possible and probably the only one that does not call for a retaining harness for the cook in rough weather. Not every yacht will have room enough for a layout of this kind and almost any other plan will demand a safety belt.

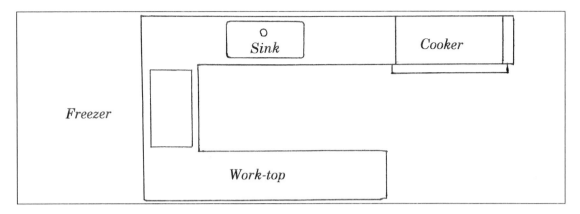

Fig 34. A recommended basic galley layout.

I think the only totally secure belt is one that straps around the waist. Sewn to this belt is a terylene strap that is free at both ends to reach out to either side. Both ends of this strap terminate in strong clip hooks that can be fastened to eye bolts set into the galley at suitable points. I must emphasis that it must be possible to strap the belt around the waist. Failing this the belt can drop down below the buttocks and then the cook can be thrown over the top of the belt in response to a sudden lurch of the boat.

A stainless steel thermos flask has proved to be of value. They are not as cheap as the plastic and glass variety but in the long run they more than pay for themselves as they are virtually unbreakable. They have one small draw back. Filled they are heavy and if they do not have a secure stowage they can move around in response to a violent movement of the boat like a misguided missile.

There are some excellent nonslip mats available nowadays that make the damp tea-towel on the galley table-top a thing of the past. Their adhesive qualities are quite remarkable. It is possible to tilt a surface on which a piece of this material is laid to about 40° without the plate sliding off. The food will fall off the plate but the plate will stay put. Certainly the mats impart a large measure of calm to the galley in a seaway.

We don't like permanent fiddles on the saloon table. They limit the usefulness of the table. Letter-writing and sundry other functions are made so uncomfortable by a fixed fiddle as to exclude the table's use for those purposes. The average fixed fiddle is usually made too low to be effective in an attempt to reduce the discomfort when the table is used for other purposes. A compromise that is probably acceptable for weekend and summer cruise type sailing but it soon becomes an irritation when the table is in daily use.

Under way we used suitably coloured bath towels as table cloths as they have a mild nonslip quality. If the movement became too lively we resorted to the clip-on fiddle I had made. As this was purpose-made I was able to make a series of partitions that would just accept our large and small plates, mugs, jam pots and the like and to make the fiddles 1½in high. By this means we were always able to take our meals at the table no matter what the weather was like.

We believe that meal times should be civilised occasions and there was no way we were going to spend several years eating and drinking from plastic utensils. Our full set of china has been in daily use for eight years now and we have experienced fewer breakages than we would expect to occur ashore.

Our stowage system for breakable china. The plates are stacked within each compartment and removed through the space at the top. Also note the high fiddle around the cooker.

Good stowage is obviously important and the picture above shows our china rack. It must be made so that the plates cannot slop about in their individual stacks. The cups are stowed in a similar manner. There are small variations in the size of table-wear items so buy your china before you decide on the dimensions of its stowage. If the fit of the stowage is sloppy you will be plagued with unceasing rattles as you roll your way to your next port of call. This applies even more so to the stowage of cutlery, but the only solution we found for this nuisance was to pack the drawer with a couple of tea-towels.

We found our pressure cooker to be a blessing. It lacks the dangers of open saucepans and economises on both fuel and water. Furthermore it is possible to cook a meal in it and leave it sealed for use later without the danger of spilling its contents. Because of the high temperature at which a pressure cooker works, the food is sterilised. If the cooker remains unopened the food can, with discretion, be kept for a little longer than food cooked in open pans and then transferred to a storage container. But do remember that the moment you open the cooker and put a spoon into it the food is no longer sterile and if you wish to keep it, the pressure must be raised again and maintained for a little while. Take a couple of rubber gaskets for your pressure cooker's lid, they are not easy to obtain further afield.

A few good quality enamelled cast-iron cooking pots and pans are by far the best type of cooking vessels. Thin aluminium pans soon become pitted and then holed and a thin-bottomed pan is for ever allowing food to burn on the rather fierce flames we are forced to use.

A measuring funnel that indicates the volume of a given weight of various materials used in cooking is worth having aboard. They do vary in the items that they cater for so check that the one you are buying suits your needs.

**Food Stores**

Since one man's meat is indeed another man's poison it would be pointless to make long lists of the food we carried but there are some comments that may well be helpful. Most places had their markets and they were often well stocked with excellent food. My mouth still waters at the thought of those luscious tropical fruits. Just occasionally you will be out of luck.

When we arrived in Nuka Hiva there were no onions available ashore until the next trading ship arrived and there were considerable shortages of a number of other items. If we had arrived three weeks earlier or two weeks later there would have been plenty of most things to buy.

Onions are so important when one is trying to make canned meat a little more attractive. My ever-resourceful wife found an American yacht at anchor in the same bay as *Didycoy* that had a good stock of onions aboard and she did a deal with the lady of the house and exchanged a dress for five large onions. There is nothing like blue-water sailing to reduce life to its fundamentals!

Rice, pulses and pasta store well and their shelf life is considerable but they are not immune to insect invasion. Good quality air-tight containers are needed for your main stock. Small containers rather than large should be used then if one container is invaded you have not lost all your stock of that item. New purchases must be treated with suspicion. Until you are satisfied that they are free of wild life don't mix them with your existing stock.

Canned foods with familiar labels may not have that familiar flavour. It seems that it is common practice to adjust the flavour to the local taste, so it pays to buy one can and try it before you buy a large number. Having decided that it is to your taste don't waste time, go and get it, in small places stocks often disappear in a day or two. One is always on the look out for canned meat. When you find some shake the can hard and listen. If the sound indicates the presence of fluid forget it, it's meat you want to buy not water! There are canned steak and kidney puddings sold in various sizes by Fray Bentos. They are excellent. We took several dozen with us. Some were used straight from the can and others were stripped of the dough and the meat was used to make chillies and the like.

We tried all the recommended methods to prolong the life-span of eggs and only one thing made a scrap of difference. They do need to be inverted about once a week. If you fail to do this the yolk eventually works its way through the albumen and ends up stuck to the inside of the shell where it proceeds to dry out. Naturally you want the freshest eggs you can get but usually you don't have much say in the matter. Eggs that are reasonably fresh will keep, unrefrigerated, for about a month or more.

All fruit and vegetables that lend themselves to the treatment should be inspected and washed as soon as they come on board and then left in the sun to dry. Bananas particularly can play host to minute cockroach youngsters. They roost at the junction of two bananas, and washing will, hopefully, remove them if they are present or at least it will draw your attention to them so that you can deliver the *coup de grace* to the little horrors.

As with eggs, we tried a variety of things to extend the life of bananas and nothing we did made the least difference. It was possible to be a little more objective with our banana tests as we were using bananas from the same hand and as I say they all ripened at the same time, no matter what we did. The myth of a stalk of bananas hanging from the backstay dies hard. No right-minded skipper would dream of tolerating that sort of weight hanging from his backstay, not to mention the state of the cockpit when all the bananas on a stalk ripened within a few days of each other!

Bruising seems to start an area of rot in fruit, so treat it all very gently. Citrus fruits appear to respond to being kept separate from each other so a little packing between them will help. When we crossed the Atlantic we took twenty seven grapefruit, one for breakfast each day as it worked out. Betty decide that the best way to keep them apart was to pack them into the legs of two pairs of panty hose with a knot between each pair of grapefruit. It worked well but to go into the forepeak where they were hanging was to witness a dance macabre, even though they were restrained by a preventer.

Tomatoes, if selected so that your stock ranges from green to ripe, will last for a week to ten days. We laid them out in egg boxes to stop them rolling about with the motion of the boat. Your big problem will be convincing the trader that you really do want some green tomatoes. When the fresh tomatoes have gone, canned tomatoes serve well in cooking with the added bonus of a glass of tomato juice for the cook or her best friend. We also enjoy the juices that come with some canned vegetables, sweetcorn, peas and carrots especially.

Fine plastic netting secured to the deckhead or a bulkhead in the odd out of the way spots makes good airy stowage for fruit and vegetables of all kinds. You must check your fruit and veg stores at least once a week and use anything that is on its way out. If you allow one item to start to decay in contact with its neighbours the rot will quickly spread.

We miss the delights of so many tropical fruits. They are as yet still unspoilt by clever agriculturists, so that the difference between what unsophisticated growers offer and what the food industry deems is good for us ... (or should that be good for them?) ... is enormous.

Vegetables that have been refrigerated before you buy them have a shorter storage life than those that have not been treated this way. This makes it difficult to predict their keeping qualities but what follows will give you an idea of the maximum and minimum periods you can hope for without refrigeration.

Potatoes will last for between three and six weeks; carrots a few days; onions for two to four weeks; white cabbage one to two weeks but cauliflower, courgettes and green beans will last for only a few days. Pumpkin and squash of various kinds will last for ages. Shop bread will last from three to seven days depending on the temperature but a fridge or freezer will extend that period considerably.

There is a belief that twice baked bread will keep longer than bread that is baked in the normal way. At Academy Bay there is a baker who will promise to double bake bread for passing yachties, at a price of course. We paid him to double bake some for us and to have it ready for the day we were leaving. Either it doesn't work or he saw us coming and failed to double bake our bread. Whatever the reason it was in line with the way we were ripped off for most things in the Galapagos Islands.

The time will come when you just have to make bread. We found dried yeast to work well and it stores well too. You will want to know how to knead the dough and this is best learned from someone who is familiar with the process. The most usual mistake made by beginners is not to really take to heart the warnings that the dough must be kept in warm, draught-free conditions if it is to rise. At first we had constant trouble with dough not rising in ambient temperatures of 90°F or more. When we realised that the problem had to be caused by the airflow through the boat, and found a draught-free place for the bread, we began to be successful.

At one time we tried a recipe for pressure cooker bread. It worked well, too well in fact. We had used our pressure cooker, after all it was pressure cooker bread, quite forgetting that our pressure cooker had a considerable turned in lip at the top. As it cooked the bread increased in volume in a most satisfactory manner until it filled the cooker and it was quite impossible to get the loaf out without cutting it to pieces. We used that recipe regularly after our initial disaster but always we used a large straight-sided saucepan. It is worth finding a recipe of this kind. Baking bread on the top of the cooker, rather than in the oven, uses less gas and does not increase the temperature of the cabin to the same extent.

If you can make bread, and you will need to, then you can also make a pizza. Bread-making days were usually pizza days for us. The yachtsman's standby ...canned tuna fish... and canned tomatoes, chopped onions and herbs topped off with some olives makes an excellent pizza filling.

American pancake mixes are on sale in many places and they turn out well. Eaten with either sweet or savoury additions they make a really nice change. Drop scones, bannocks and the like also make a pleasant change.

To keep your table salt completely dry and running free you will need a screw capped shaker. We use a discarded herb jar which has a perforated top and a good screw cap. Half fill the jar with uncooked rice and complete the filling with salt. Shake the jar until the rice and salt are mixed. Providing you always replace the screw cap your salt will stay dry indefinitely.

Some brands of margarine keep well in hot climates and do not melt in the hottest of conditions. We are fond of butter and have used canned butter to our total satisfaction. Even without a fridge we have found that it remains in good condition despite the heat. If you are running short of butter, mayonnaise is a good substitute, especially with savoury sandwiches. It also goes well with the blue-water sailor's answer to it all, peanut butter. Where would we be without our peanut butter and our tuna fish?

Australians add thinly sliced fruit to many of their savoury dishes and now that we have got used to the idea of eating a slice of fresh pineapple or mango with our fish we enjoy it.

Bean shoots have made a welcome addition to our diet on our longer passages. A variety of pulses can be made to sprout and they can make a very pleasant addition to salads and sandwiches.

Take a large jam jar, scald it and when it is cool just cover the bottom with a single layer of beans. Cover the beans with water and leave for twenty four hours. Tie a fresh piece of surgical gauze over the top of the jar and drain the water off. Cover the beans with water every twenty four hours, more often if it is very hot, and pour off the water after a few minutes. You will need to find the balance between under-watering and over-watering, if anything, over-watering is more damaging than under-watering.

The beans can be eaten when the shoots are about $1/2$ in long or you can leave them until they have grown to the top of the jar, in which case it is usual to cut the roots off before eating the shoots. Opinion varies as to whether the beans should be sprouted in the light or dark. The answer is either, but if you are growing them to long sprouts the flavour will likely be different. Our favourites are mung beans with sprouts of no more than half an inch.

We have often been asked what vitamin tablets did we take to supplement our diet? The answer is none, with all that lovely fresh food there was never a need to resort to vitamin tablets.

UHT milk and good quality dried milk is fine for cooking but we don't like the flavour it imparts to tea and so we found ourselves drinking lemon tea. At some time we were introduced to ginger tea and found it most agreeable. It is made just as you would make ordinary tea but two or three extremely thin slices of fresh ginger are put into the pot with the tea before the water is added. Most people seem to like sugar with it, even those who would not normally take sugar in their tea. We found it to be equally nice hot or cold. It is possible to use dried, ground ginger but the result is not quite as good as when fresh ginger is used.

Sea water can be added to the water used to cook vegetables to save fresh water but the extreme saltiness of sea water means that any saving is minimal. It is also possible to use sea water to shower and to wash your hair. There are some excellent detergent-based salt water 'soaps' on sale but it is necessary to rinse off with fresh water. To rinse off in sea water only succeeds in replacing the salt that you have been at pains to remove from your hair and skin.

Pots, pans and tableware washed in salt water must be rinsed in sweet water to remove the salt that would otherwise remain on them and your tea-towels and of course any clothing washed in sea water will need very thorough rinsing in fresh water before it can be worn. Bernard Moitessier recommends that clothing should be washed in salt water and then left hanging in the rigging until such time as the salt is shaken out of the garments. That may be fine for someone with the ruggedness of Moitessier, but we lesser mortals must be content with less drastic remedies.

When you are cruising, shopping often means walking considerable distances and it certainly involves carrying very large quantities of stores back to the boat. To restock a boat for the next leg can easily mean bringing back enough food for six or more weeks. It is never prudent to cater just for the period you anticipate the next passage should take. Unlike travelling by car it is not possible to pick off the distance you are to cover and then translate it into days. From Panama to the Galapagos Islands is some twelve hundred miles and in reasonable conditions it would take from ten to twelve days. We did that passage in thirty one days but friends gave up after fifty three days and went to Ecuador instead! A twelve hundred mile trip across the Indian Ocean Doldrums took us forty two days.

We found a folding shopping trolley to be the answer to our food transportation problem. With it we could easily bring back from the market at least four times the weight we could carry and with much less effort. What is more it transported many gallons of fuel and innumerable bottles of gas. Probably its greatest achievement was to transport the new wooden boom that I built to replace the one we broke in strong winds and heavy seas on our way in to Australia. We had the problem of moving this weighty, eighteen foot long spar through the streets of Brisbane. It was far too heavy for us to carry for the mile or so from the place where I had made it to the waterside. With the heavy end lashed to the trolley, the other end on my shoulder and Betty guarding the rear end we made our way back through the traffic to the river without trouble.

Chapter 20
# Security

Before we left home waters in 1983 the yachting press, and sometimes the national press, seemed to be full of stories of piracy, particularly piracy in the West Indies. Some accounts suggested that anyone venturing onto the high seas without an arsenal of lethal weapons was a fool. Betty and I debated this matter repeatedly. Every time we arrived at the same sticking point, which was to know how to distinguish between friend and foe. Do you shoot the local who has rowed out to you in some lonely anchorage? What happens if after you have shot him you discover that he has only made the pilgrimage to offer you some fresh fruit?

## An Armoury?
I was teaching navigation to a number of classes of yachtsmen and women at the time and so I offered the problem to them and also to a number of our friends. We were in a minority of two. Indeed we had not realised that we had so many blood-thirsty friends. Without exception these gentle, law abiding friends of ours agreed that we should go armed to the teeth. Some did better than that. They provided advice on the best weapons to carry and where to get them.

One student provided sketches of a string of mini-rafts that were to be towed behind *Didycoy*. Each raft was to carry a small charge of explosive ....nothing lethal you understand, just sufficient to send the enemy vessel to the bottom. It was intended that we should manoeuvre *Didycoy* so that one or more of the rafts were snugged up against the pirate ship and then we should close the switch which would detonate the appropriate charge. The following week I was presented with an improved design. In place of the switch and electric cable that was an integral part of the original model's design there was to be a small radio receiver on each raft which would detonate the charge on the receipt of an appropriate signal. Still no one could tell us how to distinguish between friend and foe and for us that was the over-riding argument.

It seemed to us, that if someone was intending to do us a mischief this fact would not be announced until the last possible moment. It was not long-range protection that we needed but close quarter defence. 'Stand by to repel borders' and all that sort of stirring stuff. I turned up a truncheon on the lathe and invested in two bottles of $CO_2$ that were intended to inflate a liferaft. The $CO_2$ bottles would throw a jet of ice cold carbon-dioxide about ten feet and we felt that a blast in the face of an attacker would at least surprise him and give us a chance to hack his shins with the truncheon. They also added to our fire fighting capacity.

We kept a white flare in the cockpit in the forlorn hope that we might indicate our position to a threatening merchantman. As a last resort we intended igniting this flare and tossing it into the pirate ship hoping that we could tiptoe away while the pirate crew tackled the blaze. Although we always put our armoury out when we were anchored in an unfrequented anchorage it was never put to the test.

When we reached the West Indies, we started to ask local yachtsmen and charter skippers who had been in the area for a number of years about piracy. They all, without exception, laughed at the idea and said it was a lot of rubbish. True, there was the occasional crime of opportunity but far less than there would be at home.

So how do these stories come about? Whilst we were still in the West Indies we received a bulletin from our Sailing Association. In it there was an account of a single-handed circumnavigation of Africa by a fellow member. Members were assured that the tales of piracy in the Red Sea seemed to be quite untrue. The account then went on to say, 'unlike the piracy rampant in the West Indies'. The writer had never sailed in the West Indies and was just repeating what he had read in the Press.

Lyn and Larry Pardy did a very careful investigation into this subject with particular emphasis on the use of fire-arms. They found that many stories were in fact fiction based on one incident that really did occur. They also looked at each case of yachties who had suffered violent death and found that all of those that could be substantiated would probably not have happened if the victim had not been carrying a gun. It seems that the production of a gun caused the other parties to anticipate the worst and react accordingly, at least one poor soul being shot with her own gun.

In the Red Sea it is not unusual for a boat that is approaching you to contain armed men. This does not indicate hostile intent since many Arabs carry arms as we carry a wristwatch. Some of the weapons carried looked so ancient that they surely would have been a serious danger to the man who dared to fire one of them. We were approached by Arabs in a number of anchorages in the course of our passage up the Red Sea and in every case we were offered gifts of fresh food and asked if we needed water. Our prime memory of the Arabs is of kindness and generosity extended to us by people, some of whom looked as if they could ill afford it.

One of the first questions one is asked when conducting the entry formalities is 'Do you have arms or ammunition on board?' If the answer is 'Yes' then the arms are taken by the Customs and kept under lock and key until the moment of your departure. At least one Asian country has adopted the routine of taking the arms and ammunition into storage and then charging for storing them. The storage charge is always greater than the value of the arms and many erstwhile armed yachties leave without their weapons.

Of course there are no-go areas and you would be unwise to ignore informed opinion but the Red Sea and the West Indies are not, at present, a cause for concern. Both areas have places that are better avoided. It would be my guess that Colombia is better missed than visited because of the violence connected with the drug traffic. Ethiopia and Eritrea have been at war for some years and I would certainly stay away from their shores. Some countries will not permit you to cruise their shoreline; Saudi Arabia was of this mind when we were in the area, so stay away. There are plenty of other places where you will be welcome.

This sort of thinking can only be commonsense, aided by information that you pick up from others who have some knowledge of the situation. Such information usually comes from two sources; yachties who have been 'there' and the boats with ham radio aboard, bless them.

Curacao in the Dutch Antilles is an interesting case. Its south shore is a succession of very lovely bays, well sheltered and with access to a road close to the shore. It also had a very high import duty on all goods brought into the island when we were there and probably still has. Northern Venezuela is but a short distance away and there was a very lucrative smuggling route from Venezuela to Curacao bringing in the more expensive everyday items thereby avoiding the import duty. The attractive bays on the southern shore were the places that were used to unload the contraband on moonless nights. It doesn't take much imagination to see that an anchored yacht in the wrong place could be considered a serious nuisance.

**Theft**

For most of our six years we left our boat unlocked when we went ashore and had no problems as a result of doing so. The only time we noticed a rowing boat was repeatedly checking us out, presumably with a view to breaking in, was in Tonga. It was dusk so I took our camera on deck and ostentatiously took a flash light photograph of them. They disappeared and were not seen in the anchorage again for the rest of our stay.

Large outboards can be a source of temptation in a few places. Its my guess that they are coveted by some local fishermen. Our scruffy two horse-power engine was never at risk and served us well, it still does for that matter. Having said that, we only heard of three outboards being stolen in six years. They were all big motors.

Chaining an outboard motor to your boat is no deterrent to a man with a pair of bolt croppers but lifting your dinghy well out of the water overnight by a halyard does make it more difficult for someone to acquire your outboard while you are asleep. It also goes a long way towards reducing the amount of marine growth that clings to the bottom of your dinghy if it is left in the water for long periods of time.

When we were there, Cristobal Colon was a lawless place and a number of yachties were mugged. By and large they were people who had disregarded advice that was freely available. Cumana in northern Venezuela was a place where it paid to be alert to the possibility of theft in the streets but it was nothing like as bad as Cristobal Colon. We sewed a pocket to the inside of the waist bands of my shorts and my wife's skirts and this simple measure gave us a great degree of comfort when we were carrying large sums of money from the bank to *Didycoy*. I made a hidey hole on board by fixing a pouch to the back of a small panel that could be unscrewed when we needed to top up our cash in hand.

The trouble with writing about this sort of subject is that it sounds like a catalogue of doom and gloom. To put things into perspective, we lost a pair of flip-flops and a whiskey bottle filled with petrol in six years. The whiskey bottle of petrol was our outboard's reserve tank. I'd not bothered to take the original labels off and I cherish the thought that a Sri Lankan docker was a sorely disappointed man!

The only incident of a serious nature occurred at dusk one day off Cape Trafalgar when we were almost home. A vessel that was behaving most suspiciously chased us for more than two hours. Fortunately we were able to outrun it although it was a tense and disturbing time. Both boats ran without lights until it became obvious that we had too great a lead to allow the pursuer to catch up with us when their navigation lights were switched on and they turned 180° and made off, eventually disappearing into the darkness. This incident reinforced my belief that no one is going to lie in wait for yachts far out to sea, they'd probably not meet up with one in a month of Sundays. Most small craft that are coasting lay their course from headland to headland and that is the obvious place to lie in wait for potential victims. Just one more argument in favour of my belief that if I can see it, I am too close!

# Chapter 21
# Charts and Books

A world circumnavigation can require three hundred or more charts, pilot charts of each ocean, and a dozen or two pilot books, radio signals books etc. If I had to economise I would cut that down to charts, pilot charts and light lists, but I would consider that to be a minimum below which I would not go.

Tide tables are not needed for most of the areas you will be visiting as tides generally are quite small by our standards, often a metre or less. In the very few places you will be staying long enough in tidal waters to need tide tables you can buy them on arrival.

References to the symbols used on the charts that you may be using will prove most useful. You may find yourself referring to charts published by national authorities who do not conform to your domestic home-water chart publisher and which may be using different scales and datums.

I found *Norries Nautical Tables* useful, often enough to be glad that I had them with me, but then I am an old fashioned navigator. But even I am now converted to using a scientific calculator in place of Norries. *Burtons Nautical Tables* are cheaper than Norries. They cover the same ground but less comprehensively.

A list of 'Lights and Fog Signals' for each area is worth its weight in gold. They make possible the safe use of outdated charts. There are a number of volumes, published by National Hydrographic Offices; each volume covers a very large area. For example, an entry for just one light in Suva Harbour will give you some idea of their value.

West Reef, south-east end.
18°08'.6S   173°23'.8E      Fl W 3 Sec. 9m high. 5 mile range.
White, round tower with a red band,
standing on a rectangular white concrete base.

Not only can you correct the light on an out-of-date chart with that entry but the description and position are of value in the hours of daylight.

On a few occasions I have found myself without a chart of a particular area and have used all the information I could muster from every possible source to make my own chart. The light list for the area was probably our most important source of information for this purpose. Your 'Light Lists' will need to be kept up to date. It would be a pointless and impossible task to attempt to keep your entire collection of Light Lists up-to-date, you will only ever need a small part of them.

I have made a practice of visiting a ship in the harbour and asking the navigating officer if I might be allowed to correct my Light List for the next leg of our passage, usually three or four pages of lights. I invariably found them to be sympathetic and willing to help.

Bear in mind that the British and American Light Lists work in totally different ways and to check one against the other is a near impossible task without making yourself a nuisance and overstaying your welcome. Because it is easy to correct your Light Lists it follows that it does not matter if your Lists are second-hand so long as they are not the copies that Noah used.

**Charts**

Both the American and British Hydrographic Offices issue Pilot Charts. For our purposes the American Pilot Charts are superior to the British versions. These American Pilot charts are produced for each ocean in the form of monthly or three monthly sheets made up into pads that cover the year. Each sheet is divided into 5° rectangles and in the centre of each rectangle there is a blue symbol that is a statement of the winds that are normally experienced for that month (or three months) their strengths, direction and frequency. Green arrows and a figure indicate the direction and speed of the current that experience has shown to be the norm for that time and place. Lines of equal magnetic variation are printed all over the Pilot Charts and these are necessary because passage charts do not normally give this information.

Recommended shipping routes are also marked on these charts and this is useful information. So far as is possible these routes are Great Circle routes. A Great Circle route is the shortest distance between two points on a sphere. For reasons of economy most ships use these routes whenever possible and so the recommended routes indicate the likelihood of a concentration of shipping. Fortunately many recommended routes cut the trade wind routes at a fairly considerable angle.

We made it a rule that when we were within sixty miles of a shipping route, or any other hazard for that matter, one of us was always on watch in the cockpit. This rule certainly saved our lives when we unexpectedly encountered a reef in the San Blas area and again in the Red Sea. I would rate Pilot Charts as essential.

Passage charts are small scale charts of oceans or parts of oceans and their adjacent coasts. They are needed for planning and for plotting your daily progress. They also make a nice record of your passages. The scale of passage charts makes them completely unsuitable for navigating your close approach to land. For this purpose you must have larger scale charts.

In a number of ways American charts are inferior to British Admiralty charts for our purposes. They are printed on inferior paper and therefore suffer from the dampness that sometimes invades a small boat's chart table. Their sizes vary which makes for awkward stowage in our limited conditions. Occasionally an area on an American chart will be left blank with a surrounding pecked line. Within the area there will be printed a warning that 'This area is dangerous to navigation and larger scale chart number ... should be consulted.' Very often it transpires that the danger to navigation is something like a four fathom patch which is no danger at all to us. The corresponding Admiralty chart would show four fathoms in the appropriate place.

If you find yourself using a French chart, check the position of the Greenwich Meridian. For some time French charts were published with the Greenwich Meridian running through Paris. The idea did not catch on and the sale of French charts slumped. As a result they returned the Greenwich Meridian to its rightful place!

To use an outdated chart in an area like the Thames Estuary would obviously be foolish but most of the places you will visit will be rocky islands, often quite steep too, their topography unchanged since Captain Cook last saw them. In places like this an outdated chart is likely to be valid, especially if it is backed up with an up-to-date Light List.

Charts of remote areas are often based on very old surveys and many such charts carry a caution to the effect that 'This island is believed to be X miles north (etc.) of its charted position'. With the precision now offered by GPS, yachties are going to find increasing numbers of discrepancies of this kind on charts that do not carry the caution. Islands can usually be seen from afar but reefs and atolls are often not apparent until you are within a mile or two of them. Differences in their charted and actual positions could be a serious hazard if you are not alert to this possibility.

A further safeguard will be your ability to pilot your vessel in relation to the colour and texture of the water's surface and this is a skill that passage makers pick-up very quickly.

Do not imagine that you can afford to go off without most of the charts you will need, intending to pick them up on the way. Your best chances lie at home where you speak the language and understand the attitudes and behaviour of your fellow countrymen.

Contrary to belief in some quarters, ships do not carry a stock of charts to hand out to passing yachtsmen. If you have a genuine problem you will find that the average ship's officer is likely to be both sympathetic and helpful, but starting out chartless hardly comes under the heading of a genuine problem. The chances of picking up more than the odd chart from fellow yachties is small, they are all going the same way as you.

Some groups of islands may have charts of the island group for sale in the harbour office of the main port. Tonga is one such group. If you can get into Nuku'alofa on Tonga'tapu, (and you will need a good chart to do that!) you can buy British Admiralty charts for the entire 150 mile long chain of islands.

If no charts are available, a friendly yachtie will almost certainly allow you to trace his chart of the area you need. When making a tracing omit the deep soundings. Put in the six and three fathom lines or the five and ten metre lines if it is a metric chart that you are using. Trace in the soundings within these lines. Add the shore line and its features and any buoyage etc that may be shown. Don't forget to be accurate on your scales of latitude and longitude.

Second-hand charts and books are often available from members of cruising organisations. A letter or telephone call to these organisations could be productive. It might pay to insert an advertisement in their club magazines.

When Chart Agents renew a ship's chart library they may have the old charts and be willing to sell them to you for a modest sum. Don't ask over the counter. Find the name of the manager and write to him, he is the person in the company who can break the rules and do something that is not within the usual routine. Make a point of reassuring the person you write to that you know the shortcomings of outdated charts and you know how to overcome the problems. Approach the Marine Manager of a shipping company in the same way but do offer to pay for the items for which you are asking. I am old enough to object to the idea of free handouts and sponsorship for people who wish to go off and enjoy themselves. If I object then it is quite possible that the recipient of your request might do so too.

It will perhaps help you to know that 300 British Admiralty charts folded to 28 x 21in need a space six inches deep.

## Plotting Sheets

Plotting sheets will be needed for a number of reasons. They can be difficult to find and when you do find them they usually prove to be expensive. There is no need to buy them. If you are working within 15° of the equator, the length of one degree of latitude is so close to one degree of longitude that it is reasonable to use squared graph paper as your plotting sheet. When you have done your work on the plotting sheet you can then transfer the position to the passage chart. Even at 20° north or south of the equator the error is so small as to be acceptable on the open ocean, it's probably not measurable on the small scale of a passage chart.

When you wish to plot beyond these latitudes, it is simple enough to make your own plotting sheets. I use graph paper when I am constructing a plotting sheet simply because my log book consists of alternate lined and graph pages but there is no reason why they cannot be made on plain paper.

Draw a vertical line down one side of the paper and at 90° to it draw a line close to the bottom of the sheet. Mark the base line in equal divisions; for this exercise use one inch divisions and label them as 10' of longitude. (Figure 35)

From the junction of the two lines add a further line so that it makes the same angle with the base line as the latitude in which you wish to plot; for this exercise make it 40°.

From the first of the ten minute longitude divisions raise a perpendicular line until it reaches the diagonal. (Figure 35)

With your dividers measure the distance from the junction of the vertical and horizontal lines to the point at which the perpendicular line meets the diagonal line. This will be the length of ten minutes of latitude and should be transferred to the vertical line to give you ten minutes of scale all the way up the line. (Figure 35)

The choice of one inch for the longitude scale is not mandatory, any scale may be used depending on the purpose to which you wish to put the plotting sheet. Often, with a sheet of graph paper, it is possible to use the divisions that are printed across the paper. There are mathematical ways to produce a plotting sheet for a given latitude but they are all much more complicated and for all practical purposes no more accurate.

It is possible to copy the scale of a chart of the latitude you wish to use for your plotting sheet but it will almost certainly be the wrong scale for your purpose.

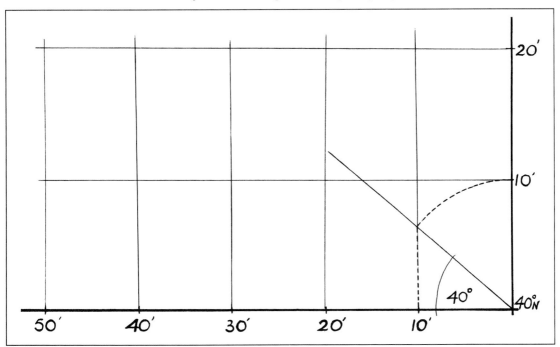

Fig. 35

# Chapter 22

# The Compass

It is essential that you understand the causes of compass deviation, how to detect and measure it, how to reduce it and how to allow for it in your navigation. None of these matters are particularly difficult to understand.

## Deviation
Anything that creates a magnetic field within range of your compass can cause the compass card to deflect from its true position. Electrical circuits and ferrous metals are the prime villains. Soft iron, the most common form of iron used in everyday things, will become a temporary magnet when placed in a magnetic field. Since it is always in the earth's magnetic field it is always behaving as a weak magnet. Because of the temporary nature of its magnetism, soft iron will change its polarity if its alignment within the magnetic field is changed. (Fig.36)

It is most important that you understand and remember that last paragraph because it is the key to the nature of deviation.

Hard iron, once magnetised, remains a magnet and unlike soft iron it does not change its polarity, but the amount of hard iron to be found in a boat will be small compared with the volume of soft iron. The compass needle is one example of the use to which hard iron is put. Although the amount of hard iron to be found in a yacht is small its effect cannot be ignored. Because soft iron is likely to be the greatest deviating factor on board it follows that your vessel's deviation will change with each course.

Figure 37 shows the same boat on four different headings. In each case the rectangle represents the total of the deviating forces acting on the compass, most of which is soft iron that changes its polarity with each change of course. The circle represents the compass and shows how the compass needle, which is free to move, responds to the changed polarity.

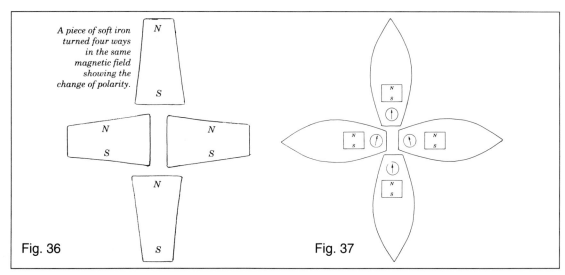

A piece of soft iron turned four ways in the same magnetic field showing the change of polarity.

Fig. 36

Fig. 37

With a direct current such as we use aboard our boats, electrical circuits and wires have their own magnetic fields as soon as the power is switched on. The magnetic field around a wire carrying DC is arranged in a specific pattern causing that wire to become a magnet. If both positive and negative wires are run side by side, the pattern of the magnetic field in one wire will cancel the pattern surrounding the other. Therein lies your defence. Never run a single wire in your boat, always run positive and negative wires together.

Soft iron comes in many disguises ranging from engines to anchors, baked bean cans and tin openers. Much of it is moveable so common sense stowage is required to avoid problems. Don't forget that magnetism passes through bulkheads as if they did not exist, so take care what you fix to the bulkhead behind the compass.

Magnetic power diminishes as the inverse square of the distance. This means simply that the deviating effect does not fall off as in a straight line graph but it reduces more and more rapidly as the distance is increased. In practical terms this means that if a piece of gear is causing a problem, moving it a short distance away may be more effective than the distance might lead you to expect.

Many pieces of electrical equipment will develop magnetic fields when they are switched on. Some will retain their magnetic field when switched off. Your only defence here is to site items of this sort far enough from your compass to be sure that they cannot influence its behaviour. Three feet is the usually accepted minimum distance.

Not all grades of stainless steel are non-magnetic. If in doubt check with a magnet or hold it close to the compass in three or four positions and watch for a movement of the needle as the stainless steel approaches the compass. The only metals you can be sure of are the nonferrous metals, brass, bronze, copper and aluminium. Objects like cameras, binoculars and knifes may well cause the compass to give an inaccurate reading if they are stowed too close to it.

Compasses in steel or concrete vessels are obviously exposed to considerable deviating influences caused by the steel used in the construction of the vessel. There are three possible solutions to this problem. If your cockpit is open and big enough to accommodate a binnacle, a large non-magnetic support that will place the compass at least three feet from any source of deviation, this will reduce the problem considerably. I have seen a number of steel yachts successfully set up in this fashion.

It is possible to buy a compass with a good correcting system built into it. You can then site the compass in the best possible position, swing the compass to find its deviation, correct and swing again until you are satisfied that you have reduced the error to manageable proportions and also know what deviation remains for each course. There are also compasses that are designed to be fixed to the mast beyond the influence of the hull etc that will operate an electrical repeater in the cockpit.

If you are forced to adopt either the second or third solution I think you would be wise to deal direct with the technical department of the compass manufacturer rather than a chandler. Most chandlers will do their best for you but what you need is expert advice and clearly the chandler cannot match the expertise of the manufacturer.

In case you are not totally familiar with compass terms it may be as well to define the few we need to use.

**True North** is the direction taken by a line extending from any point on the earth's surface and the geographical north pole. A 'True' direction will be related directly to True North.

**Magnetic North** is the direction taken by a line extending from any point on the earth's surface to the magnetic north pole. The magnetic north pole is situated somewhere in northern North America.

A 'Magnetic' direction will be measured from the magnetic pole.

## Variation

The angle between True North and Magnetic North is called Variation and it may be either East or West. Variation changes a minute amount each year and its value also differs in various places on the earth's surface.

**Compass North** is North as indicated by a compass needle. If there are no outside influences Compass North will coincide with Magnetic North.

## Deviation

In a vessel the compass needle will normally be subjected to interference from the structure and contents of the boat and this error is called Deviation. Like Variation it may be East or West.

## To Swing a Compass

To swing a yacht's main steering compass is a fairly simple operation and well within the competence of the average yachtie. If you are to take off into the blue then I think it rates as one of the many minor skills you should possess.

What is needed is that you should be able to take bearings of a distant fixed mark with your steering compass in its working position, whilst the boat is held on each of the eight cardinal and intercardinal points. It does not matter that the position of the mark is unknown but it is important that it is a long way from your boat. This last requirement makes for greater accuracy.

In many yachts the position of the steering compass will make it impossible to take bearings with it. This problem can be overcome, but for simplicity's sake we will talk as if it is possible to use your compass in this way. Later, if your compass is badly sited for this purpose, your special needs will be catered for. Whatever you do, don't move your compass to a temporary position where you can take bearings with it. Any deviation card calculated in this way would only apply to your compass in that position.

If you are to swing in a tidal area you must work at slack water otherwise you will find it difficult, or even impossible to manoeuvre the boat. For the same reason you must choose a day with little or no wind. Three people will be needed; one to run the outboard dinghy that will push the yacht around its anchor, one to handle the helm and to state when the boat is on the required heading and a third to take and note the bearings.

With your yacht anchored in a position where you can see your distant fixed mark you are ready to begin work. Remember, the aim is to obtain the bearing of your chosen mark when the boat is on each of the eight cardinal and intercardinal points. Record each of the eight bearings in the way they are set out in the column headed 'Ship's head by compass'. For the moment ignore the remaining two columns in the example shown in Figure 38.

When the eight bearings are recorded they must be totalled and divided by eight to find the average of all the bearings.

Deviations, when plotted as a graph, usually fall into the pattern of a sine or cosine curve. As you can see from Figure 39, the values on one side of the centre line equal the

| Ship's head[1] by compass | Compass[2] bearing | Magnetic[3] bearing | Devn[4] |
|---|---|---|---|
| N | 125° | 125° | 0° |
| NE | 131° | 125° | 6°W |
| E | 133° | 125° | 8°W |
| SE | 131° | 125° | 6°W |
| S | 125° | 125° | 0° |
| SW | 119° | 125° | 6°E |
| W | 117° | 125° | 8°E |
| NW | 119° | 125° | 6°E |

|  | 1000° | ÷8 |
|---|---|---|
| = | 125° | Magnetic bearing |

Fig 38.

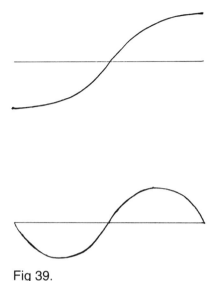

Fig 39.

values of those on the other side. From this it follows that the average of the eight bearings of your distant object must be the one that is free of deviation since the deviations on one side of the centre line must cancel the deviations on the other. This bearing is both a magnetic bearing and a compass bearing. All those affected by deviation are known as compass bearings.

The difference between the compass bearings in Figure 38 and the magnetic bearing is the value of the deviation. As you can see the deviation for North (Ship's head by compass) is nil but for South-East it is six degrees. It remains to name the deviations East or West.

There are a number of ways to remember how to decide whether a deviation should be labelled East or West. We will use the jingle 'deviation east, compass least'. In Figure 38 the compass bearing for South-West is 119° and the magnetic bearing is 125°. Clearly the compass bearing is the least, i.e. smaller, which means that the six degrees difference must be easterly deviation.

The same applies to the deviations for West and North-West. From this it follows that the deviations for North-East, East and South-East must be westerly deviations.

It may be that, like me, you prefer a visual presentation and so Figure 40 may prove to be more to your liking. The larger circle is the 'magnetic circle' and the smaller circle represents the compass needle. In the example we require the vessel to sail on a magnetic course of East. You can see that when the compass needle is deviated to the East then the compass course to steer will be less than 090°, 'deviation East, compass least'.

If you need to experiment further to convince yourself, then cut a circle of card to represent the compass-card and fix it above the larger circle with a drawing pin and play with it.

When you feel happy about naming the deviations East or West, plot the deviations against their compass headings as in Figure 41. The curves should be smooth. A bearing that does not conform to the curve should be treated with suspicion and, if need be, taken again. When that is completed intermediate figures can be added as in Figure 42.

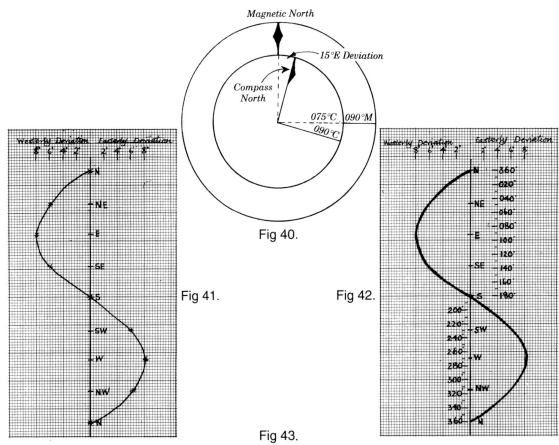

Fig 40.

Fig 41.

Fig 42.

Fig 43.

| Ship's head by compass | Devn | Ship's head Magnetic | Ship's head by compass | Devn | Ship's head Magnetic |
|---|---|---|---|---|---|
| 000° | 0° | 000° | 190° | 2°E | 192° |
| 010° | 2°W | 008° | 200° | 3°E | 203° |
| 020° | 3°W | 017° | 210° | 4°E | 214° |
| 030° | 4°W | 026° | 220° | 5°E | 225° |
| 040° | 5°W | 035° | 230° | 6°E | 236° |
| 050° | 6°W | 044° | 240° | 7°E | 247° |
| 060° | 7°W | 053° | 250° | 7°E | 257° |
| 070° | 7°W | 063° | 260° | 8°E | 268° |
| 080° | 8°W | 072° | 270° | 8°E | 278° |
| 090° | 8°W | 082° | 280° | 8°E | 288° |
| 100° | 8°W | 092° | 290° | 7°E | 297° |
| 110° | 7°W | 103° | 300° | 7°E | 307° |
| 120° | 7°W | 113° | 310° | 6°E | 316° |
| 130° | 6°W | 124° | 320° | 5°E | 325° |
| 140° | 5°W | 135° | 330° | 4°E | 334° |
| 150° | 4°W | 146° | 340° | 3°E | 343° |
| 160° | 3°W | 157° | 350° | 2°E | 352° |
| 170° | 2°W | 168° | 360° | 0° | 360° |
| 180° | 0° | 180° | | | |

Figure 43 shows the finished job with the figures set out for easy selection. Note that the total easterly deviations exactly equal the total westerly deviations. This is the most usual result. If one set of deviations had been greater than the other, assuming an accurate swing, there are three possible causes. A broken compass pivot point could cause this problem but it is the least likely of the three possibilities. A more common reason is that the main mass of your deviating influence is to one side of your compass. But the most usual cause of this form of error in a yacht is mis-alignment of the lubber line. If the lubber line (the fore and aft line of the compass) is not aligned with the fore and aft line of the boat then this kind of error will make itself apparent.

If you do a swing in a steel or ferrocement boat the results may not conform to the tidy set of curves that one usually gets in a timber or glass fibre boat.

A compass with built-in correctors, or a binnacle with the necessary set-up, will allow you to go a long way towards removing most of the deviation that might be found even in a steel boat. If you can see a likely reason for the odd deviations it may be possible to remove the offending piece of equipment or even resite the compass. In the event that you can do nothing to improve matters at least you will know just what the deviation is on all points of sailing.

It is not generally appreciated that a major change of latitude can cause changes in a vessel's deviation, so it is as well to check the performance of the compass whenever possible. Every time that you enter a harbour that has a set of leading marks to guide you in, you can check your compass on the heading of the leading marks and when you leave you can check it on the reciprocal of that heading. There is no need to do the calculation there and then. So long as you write down your compass course when you have the marks in line you can do the work later.

For example:-
    The leading marks into Townsville bear 212°T from seaward.
    The variation for that area is 8°E which makes the heading 204°M.
    If your compass reading is something else the difference must be the deviation on that course.

At dusk and dawn every day a compass check is possible for the course on which you are sailing. The point on the horizon at which the sun rises or sets is somewhat to the North or South of East and West according to the declination of the sun and your latitude. The declination is taken from the current Nautical Almanac.

There are tables in *Norries Nautical Tables* that will solve the problem for you quite simply. *Norries* are a rarity in yachts today but a scientific calculator will do the same job for you just as easily and solve your astro problems too.

To work an amplitude by calculator you will need to use the following formula.

$$\frac{\text{Sine declination}}{\text{Cosine latitude}} = \text{Sine Amplitude.}$$

Calculators are unable to digest a formula presented to them in this fashion but if you turn it around to read 'Declination Sine ÷ Latitude Cosine' the calculator will show its gratitude by giving you an answer that is the sine of the amplitude. Press the Second Function key ( 2F ) or in some models the INV key, and then the Sin key, and the sine of the amplitude will be changed to degrees which is what we want.

Two more simple steps will complete the operation. The amplitude must be named and then it must be converted to 360° notation. To name the amplitude it is first called East for a rising sun or West if it is setting and this is followed by the same name as the declination i.e. North or South. Some examples should make this clear.

Example 1:-
    The calculator gives a sunset Amplitude of 12°,   Declination is North 15°.
    The Amplitude is named West 12° North.
    West is 270°T and 12° North of West is 282°T.

Example 2:-
    The calculator gives a sunrise amplitude of 18°,   Declination is South 5°.
    The Amplitude is named East 18° South.
    East is 090°T and 18° South of east is 108°T.

The bearings arrived at in Examples 1 and 2 are the bearings at which the sun could be expected to set or rise on those two occasions. If you took a bearing of the sun at the appropriate time with your steering compass and then compared it with the sun's amplitude, converted to a magnetic bearing by the addition or subtraction of the local variation, the difference would be the deviation for the course you were steering at that moment.

Example 3
    At sunset the sun bears 295°C (by steering compass). Variation 8°W
    Declination North 10° Latitude 20° North.
    What is the deviation for this compass course?

    The calculator gives an Amplitude of 10°.6, let us call it 11°.
    When the Amp. is named it becomes W11°N which equals 270°+11° = 281° the true bearing of the setting sun.

| | |
|---|---|
| Compass bearing of the sun | 295°C |
| True bearing of the sun by Amp. | 281°T |
| Compass error | 14°W (Compass best, error West) |

**Compass Error**
Compass Error is the sum of deviation and variation. If we have 14°W compass error and 8°W of it is variation the remainder of 6°W must be the deviation. (Fig 44)

To help you relate to the three sketches remember that compass error and variation are measured against True North.

If the variation had been 8°E with the same compass error the deviation would be 22°W, (Figure 45).

With a variation of 18°W and a Compass error of 14°W the deviation would be 4°E, (Figure 46).

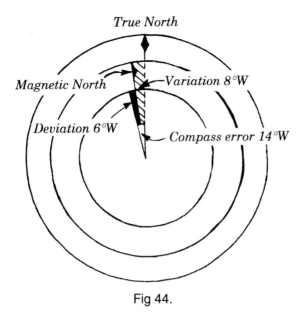

Fig 44.

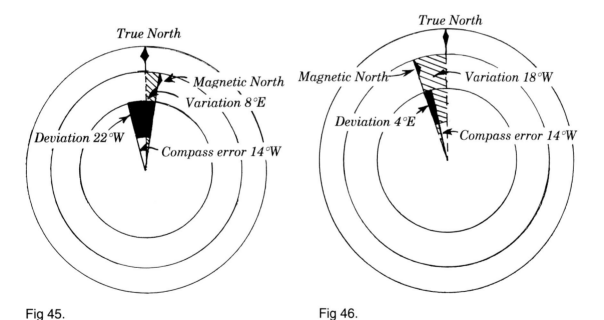

Fig 45.                            Fig 46.

The correct moment to take an amplitude of the sun is when its lower limb, the bottom edge, is a semi-diameter above the horizon. Refraction bends light rays around the curvature of the earth and it is necessary to counter this by taking the bearing at this time. Other bodies can be used for amplitudes but the sun is the one most often used.

**The Pelorus**

If your compass is sited so that you are not able to take bearings with it, you will need to make a pelorus. The pelorus is simply a dumb compass. In use it is set to the same heading as the steering compass and then when you take a bearing with it, it will read the same as if you had taken the bearing with your steering compass.

To construct a pelorus you will need a circular plastic protractor, a square of plywood that is a little larger than the protractor, a home-made sighting vane, four small rubber-suction cups, a nut and bolt and four small screws. The plywood, the protractor and the sighting vane are each drilled through their centres with a hole that will just accept the bolt. Hardware stores sell small rubber suction cups with a hook screwed into them. They are intended to cling to a tiled wall to provide somewhere to hang a tea-towel. Four of these with the hooks removed will make the feet for your pelorus, screw one to each corner of the plywood.

Draw a line across the ply so that it crosses the centre and is at ninety degrees to one side. Bolt the vane, protractor and the ply together and the pelorus is ready for use. Treat the line as if it were the lubber line of your compass.

When you want to use the pelorus, set it up on the hatch top or some similar place where you have a clear view. Align it carefully with the vessel's fore and aft line and set the protractor to the compass course that is being steered. So long as the boat is held on course, any bearing taken with the pelorus will be the same as if you had taken it with the steering compass. Not only will this allow you to swing the compass, but it will also serve in place of a hand bearing-compass if need be and allow you to take amplitudes.

When it is being used as a hand bearing-compass, it will be important to remember that any bearings you may take with the pelorus will need to have the deviation for the course being steered applied to them. This does not apply when you are swinging the compass or using it to take an amplitude.

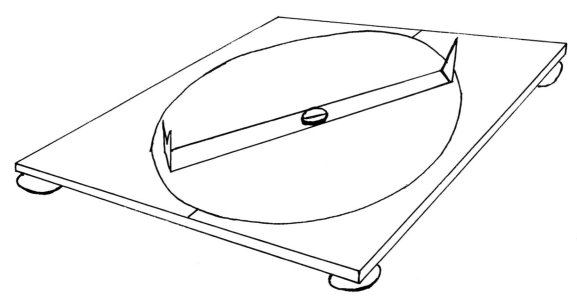

Fig 47. A Pelorus

**Compass Care**
Strong tropical sunshine over a period of time, can damage the transparent plastic that forms the face of the compass. If your compass is in an exposed position then you must either cover it or take it below whilst you are in harbour. I stumbled over this problem in Brisbane. A fellow yachtie was complaining that he had to buy a new compass because the plastic face had become corrugated under the influence of prolonged exposure to strong sunlight and he could no longer see the card clearly.

I suggested that he had very little to lose by trying to correct the situation for himself and perhaps the price of a new compass could be the prize for success. Some wet or dry abrasive paper was used to get rid of the corrugations that had appeared on the surface of the plastic face. By using finer and finer grades of wet or dry abrasive paper and then finally polishing with a metal polish for some time, he produced an excellent finish. The whole job took less than a day to complete and he rated it as the most profitable day's work he had done in a very long time!

There is an understandable tendency on the part of many yachties to feel that a compass is something that is outside their sphere of competence. If you are to sail in distant waters you will need to undertake the care and maintenance of many items that, at home, you would hand to a professional to put right, and a compass is one of them. Of course a compass must be handled with care but it is not quite the fragile thing some people seem to believe it to be.

I met a Swedish yachtie who was bewailing the fact that he was going to have to buy a new compass because he had been quoted $500 and a six week delay to have a bubble taken out of his compass. (At $500 Aus. for an hour's simple work he was being taken for a ride. Presumably the six week delay was to justify the $500.) He was reluctant to believe that we could do the job ourselves but eventually agreed to give it a go. Parting with $500 was painful enough but the six weeks delay was intolerable, as it would put him into the wrong season for the next leg of his voyage.

Let me say loud and clear that a bubble in a compass does not effect its efficiency in any way. It may be an irritation and if the situation is not rectified for a very long time the bubble could reach a size when there was not enough fluid left to dampen the movement of the card in a seaway, but it would take a long time to reach that stage.

On the side of your compass you will find a large grub-screw which is fixed into an upstanding collar. Turn the compass to put this screw uppermost and support the compass securely in this position. This is a task to be performed at anchor, not at sea because you need a stable situation in which to work. Remove the screw and lodge it in a safe place. With a hypodermic syringe and needle remove a little of the fluid for examination. It will be either oil, in which case you will be able to feel the oiliness when you rub it between your fingers; alcohol, your sense of smell will help and the fluid will feel cold as it evaporates on your fingers; or a fluid that seems to be closely related to, or maybe actually is, white spirit, if you have ever used it to thin paint or to clean paint brushes you will recognise it at once.

Replace lost white spirit with white spirit, alcohol with gin or vodka ....you cannot buy the unadulterated alcohol used by compass technicians without a licence .... and the oil can be replaced with Johnson's Baby Oil. I have used all three on compasses of my own with no ill effects. Just make sure you have the right fluid for your compass, mixing fluids could bring problems. A syringe and needle and a lot of patience is all that is needed. The difficulty is getting rid of the last bit of the bubble. Tapping the side of the bowl and moving the bowl very slightly can help, but mostly it is patience that is needed.

A bubble in the compass fluid obviously means that fluid is leaking out and air is taking its place, albeit very slowly. There should be a neoprene washer under the grub screw and this is the most likely source of your leak. Ideally it ought to be replaced and if you have some neoprene sheeting about 3 to 5 millimetres thick you could cut one to fit. The hole for the screw can be drilled before you cut the washer if you don't possess a suitable sized punch.

If the bubble reappears and you are satisfied that you have done a satisfactory job of sealing the screw then you must look elsewhere for the leak. Almost certainly it will be at the junction of the two halves of the body of the compass or the face plate if it is that type of compass. This will require a new gasket and is a bigger but no more difficult task than removing a bubble, but don't rush into it before you are certain that the washer under the screw is not to blame. I suggest that you work on your compass in a plastic washing up bowl, you tend to drop fewer minute screws on the floor that way.

## The Compass-card

The compass-card has developed over many years to its present form. Not surprisingly the early cards were divided into 90° sectors and these were then subdivided. This served well enough for a long time and indeed the cards of this time with their central feature of a large star were very comfortable cards to steer by.

With the card divided into thirty two 'points' a point became $11\frac{1}{4}°$ which was bad enough, but further divisions were made and the system eventually became too unwieldy to use and a change to the present system of degrees became inevitable.

Weather forecasts, pilot books and the like still use the system, usually when the author wishes to indicate an approximate direction. An illustration of a card based on the points method of notation is included to help you should you need to decipher instructions given in this form.

Fig 48

Fig 48. The traditional compass card.

Chapter 23
# Finance

'What does it cost?' Obviously it is not possible to give a specific answer in precise figures. So many factors rule that out, but it is possible to give some indication of likely expenses.

The cost of your boat and fitting it out will be, to a large extent, a personal choice. Once that is settled there will be ongoing expenses and that is the area where useful comment can be made.

If you wish to eat at the same standard as you do at home, on average it will cost about the same wherever you are. There will be places where fresh food will be cheap but canned goods seldom are and you will need a large stock of canned food to see you safely through the longer passages. Many of the smaller islands in the Pacific have much of their food brought to them by ship which adds rather to the cost.

Tropical fruits are mouth-wateringly delicious and often ridiculously cheap but whilst you can, and doubtless will gorge yourself on them, you cannot live on fruit alone. If you can sharpen your fishing skills and you have a freezer on board in which to store part of your catch, some saving is possible but don't rely on it.

Repairs, anti-fouling, postage, launderettes and laundries, if you can find them, all cost about what you would expect to pay at home. Chandlery can be expensive and difficult to find. Diesel was always cheaper than at home, sometimes unbelievably cheap, and a haul-out was often 25 per cent cheaper than at home.

A limited number of countries make a charge of some kind when you enter or leave their territory. The charge goes under a variety of names .... visa, harbour dues, light dues, exit tax and so on. In the six years from 1983 to 1989 these dues cost us about £200, a mere £33 a year. By avoiding Sri Lanka, Venezuela and the Galapagos we could have reduced that figure to £80 or £13.30 a year.

In the same period we entered eleven marinas but that could have been reduced to two if we had wanted to save money. For the rest of the time we anchored and that cost nothing. The period at anchor would vary from overnight to several weeks. Many of our anchorages were so lovely that I still wonder how we tore ourselves away from them. When I read of the people who rush around the world to make the voyage faster than the last boat I think, 'Poor souls .... how sad to be so afflicted!'

Beware of buying a boat abroad, perhaps to save money. You will almost certainly be charged import duty when you return home. Regulations change it is true, but this duty can be a substantial sum so at least check with the Customs Authority before you commit yourself to a purchase.

We relied on credit cards for our answer to the problem of drawing money in most places around the world. I must say that once we had sorted out the teething troubles it was excellent. If you are not familiar with the system it operates like this. The credit card company sets a maximum sum that they are prepared to let you owe them at any one time. (This obviously relates to your financial standing in their eyes.) Some weeks after you have made a withdrawal, they will send you a bill asking you to repay them. Should you fail to do so on time they will start to charge you interest on the money outstanding. We arranged for all our card bills to be sent to our bank who then paid them out of our funds. On the whole this arrangement worked well, but at first our withdrawal limit was set too low, not in relation to our income but in relation to the speed of international communications.

In this age of satellites, telex, fax and goodness knows what else, it seems incredible that it sometimes took weeks for the information to travel from our point of withdrawal to company headquarters in England. Incredible or not, that was what happened and we were constantly being told that we were out of credit. When you apply for a withdrawal, the bank you are dealing with will telex the local credit card company headquarters to establish the state of your credit-worthiness. This can take a few minutes to several hours in some places.

When we informed our bank of the problem, the card company raised our cash limit considerably and we have had no trouble since. The same card can, of course, be used in a great many shops and stores and this can be a useful facility. A card also gives you the ability to pay for goods supplied by some companies at home. When you write or telex the order to them, you simply include your card number and ask them to accept payment by this method. Obviously you must know in advance if that company is willing to accept payment by that brand of card.

If two of you are applying for cards, do get them at different times. If both your cards expire at the same time you could be cashless for too long. Before you leave home, visit your bank manager and put him completely in the picture. We did and they were most helpful. Get a telex or fax number from him, then if the need arises you will be able to contact your bank as quickly as by telephone and often very much more cheaply.

The telex and fax numbers and the addresses of companies that may be able to provide spares and advice for you are also worth collecting.

Establish a method of transferring money with your bank manager. When we were in Australia the crankshaft of our engine broke. The cost of the spare parts from home that we needed, plus air freight etc was £759. Unfortunately this company did not accept credit cards. To get the money from our bank in England to the spare parts supplier, also in England, it went something like this.

Sterling from England was changed to Australian dollars so that we could withdraw it from the Australian bank with our card. In the same Australian bank, but at a different counter of course, I had to change the Aussie money to American dollars so that it could be sent to England. Naturally it didn't go to our bank but to the Australian bank's British agent bank so that they could change the U.S. dollars to Sterling and then send what was left of it to our bank. At each stage our money was well and truly milked until that little bit of bankers' snakes and ladders cost us well over £100 to send our £759 from our bank in London to the spare parts company in Coventry.

Brisbane was our nearest major airport and the company in England consigned our spares to Bowen, via Brisbane Airport. To quote from our journal.

> '*At the time of writing we know that our spares are in a sealed container in Sydney Airport, a mere 600 mile navigational error. The Customs, for whom I have nothing but praise and gratitude, have found the package for us but are not allowed to see it until the freight company decides to open the sealed container in their presence. I am told that it is not unknown for a shipping agent to hang on to a consignment so that they can charge for storage. (We were later required to pay $14 for storage at Brisbane, an airport our package never reached, before it could start its journey from Sydney.)*

*Meanwhile we sit here paying harbour dues and growing barnacles on our bottom, with our Australian visas running out and the cyclone season upon us.'*

Throughout the South Pacific it was uncommon to find a bank that would handle credit cards of any kind except in the bigger islands like Tahiti. We knew in advance that this would be so and had laid in a good stock of American dollars, which seemed to be acceptable almost everywhere. We knew in the same way, as we all knew so much about conditions ahead .... by courtesy of the ham radio enthusiasts.

## Mail

It is difficult to offer much help on mail. You will need an agent at home who will accept mail for you and then send it on when you have settled somewhere for two or three weeks. If you can find a fax office, that is probably the quickest way to inform your agent of your arrival. A letter that is typed or handwritten is fed into the machine and then photocopied and transmitted to its destination within minutes. If your agent does not have a fax number, then send it to the agent's address c/o the post office and they will post it the next day by normal post.

French 'post restante' services will hold mail for two weeks and then return it. Australia and New Zealand both have excellent mail services with a variety of forwarding arrangements. The Frangipani Hotel, Bequia, British West Indies and Don Windsor, Galle, Sri Lanka will both hold mail indefinitely for yachties, but do write and let Don know that you are on your way. When we were in the Sudan there was no mail or telegraph communication with the outside world. I suspect that poor Sudan has more immediate things on which to spend what little money it has. South American mail was unreliable to say the least.

A few countries require you to have a visa before arriving in their territory. This is usually common knowledge well in advance, courtesy of the ham radio network. Australia requires all members of the crew, except the skipper, to arrive with a visa. The skipper gets a cruising permit free of charge.

## Work Permits

Most countries are unwilling to let visiting yachties work for reward, the exception being those who perhaps have a skill that is in short supply ashore when a work permit may be issued. This rule is enforced with varying degrees of stringency from one place to another. In the Galapagos I was required to buy a work permit to allow me to scrub the bottom of our boat and I had to pay 10 Sucres for the permit to be typed!

It is possible to earn a little money by working for other yachties but a specialist skill is needed. Engineering, engine, sail, electrical and electronic repairs are the skills that are most often wanted with hairdressing trailing further down the scale. None of these jobs will bring in enough money for you to live on unless you are prepared to stay in a place that is a hive of yachting activity and where the authorities are willing to turn a blind eye to your activities. If you can find such a place and you are able to stay long enough to accumulate sufficient cash to see you through the next leg of your voyage you will have done well.

Don't imagine that you will be able to support yourself by doing any work, no matter how menial. There are many thousands of people in the third world countries who would be delighted to have the chance to do just that.

Magazine articles are really not worth the effort as a general rule. What blue-water sailors are doing is not what excites editors at home unfortunately. Perhaps if your name is well-known in yachting circles it might help, but the competition must be considerable and the opportunities limited. The only person we met who was making a living from writing about her sailing was a Canadian journalist. She was an established professional in her own country and employed an agent to get her articles published. I might say she worked very hard at it too.

## Insurance

Insurance is a very difficult area for passage makers and I cannot understand why it should be so. The insurance companies compete with each other to insure those who race and doubtless provide those company's claims departments with most of their work. Those who pretended to consider us insurable asked for annual premiums that could have been mistaken for international telephone numbers.

In desperation I suggested to one well-known company that perhaps the premium could be lowered if the policy was only valid when we were within fifty miles of land. They replied with some show of interest saying that it would be possible on the following conditions.

> 'When we arrived somewhere we should haul-out and have the boat surveyed. When they received and approved of the surveyor's report they would write back to us and tell us that we were insured.'

For most of our circumnavigation haul-out facilities were few and far between and surveyors were as thick on the ground as hen's teeth. The world's postal system being somewhat less than perfect it could easily have taken several weeks to get a letter home and a reply back to us. Meanwhile we would have sailed on, uninsured, in search of another slip and surveyor to relieve us of yet another £500.

This was the reply that convinced us that the insurance companies knew less about the conditions we were about to face than we did. We were to discover that the vast majority of passage makers are uninsured. Once we became accustomed to the idea that we could go to sea without insurance we found that it was quite painless. If most blue-water sailors are uninsured, I wonder where the insurance companies get their statistics from to back their claims that our form of sailing is so very dangerous?

In all my years of sailing I have never before mixed with such a competent and mutually supportive group of seamen so devoted to the welfare of their vessels. One hears of incredibly few incidents that would involve insurance claims.

## Health Insurance

Health insurance was no better. Naturally no insurance company would be so foolish as to insure you against the worsening of an existing condition .... it might happen! Most companies considered us to be too old to insure anyway. The best offer we received had to be renewed every three months and to do this it would have been necessary for both of us to return home every three months.

**Extraordinary Cost**

Tahiti is an immensely popular crossroads for boats in the South Pacific. At a guess I would say that there could be at least 200 boats there at any one time. It is usually just outside the eastern edge of the cyclone area so a great many yachts sit out the cyclone season there with many more scattered around the Tuamotus and the Marquesas. This is all French territory and the French authorities require each yachtie to deposit a bond for the duration of his stay. The bond is returned in full when they leave the area. The size of the bond is determined by the cost of the air fare back to the depositor's native land. As you can see this is a substantial sum of money and a measure of resentment arises because different nationalities seem to be treated differently. This is not a matter of discrimination, it simply reflects the cost of repatriating people to different parts of the world.

Middle Percy Island — Great Barrier Reef.

# Chapter 24
# The Panama and Suez Canals

**The Panama Canal**

We had heard numerous horror stories concerning the difficulties and dangers inherent in a passage through this wonderful waterway and I am happy to report that they are mostly either figments of the imagination or gross exaggerations. There are difficulties of course but they are almost entirely due to the turbulence in the three up-locks, particularly the first one. The three down-locks have currents running in them when the gates are opened to let you out, but if you follow the pilot's instructions the problems are usually overcome without too much pain.

All the way from Europe we were told repeatedly and with great certainty that it would cost $1,000 for us to pass through the canal, plus a substantial measuring fee. One boat yard owner in the West Indies actually claimed to have telephoned the Panama Canal Company to get the latest information for the benefit of passing yachties. According to him the Canal Company had confirmed that $1,000 was the current charge and was to be raised in January 1985 by an unspecified amount.

In April 1985 we were charged a total of $105 (US) for the passage of a 36ft Hillyard. Ninety days later we were sent a refund of $25 'Because we had not damaged the canal'! $80 was the cost of measurement, use of the canal and locks complete with a very competent 'Adviser' (Pilot) in attendance for rather more than 12 hours. In this crazy world you will not be surprised to learn that of the $80, $17 was for the use of the canal and the services of the Pilot and the remaining $63 was for the paper work!

Your first task in Cristobal Colon, or Balboa if you are coming from the West, is to clear in with Customs and Immigration. The next step is to get your 'Cruising Permit'. Don't attempt to do this at the weekend, it will cost you double the week-day price to cover the cost of 'overtime' work. When this is done you will be free to ask at the 'Admeasurer's' office to have your vessel 'measured' as the first step in organising your transit.

Sometimes the same day, sometimes a few days later, depending on how busy they are, an official will arrive clutching a large sheath of forms and a tape measure. It took us about forty-five minutes to complete the paper work which, as ever, is designed for big ships, not for yachts. I signed several times as Master, Chief Engineer and finally as ship's Surgeon.

Four or five measurements were taken from which the fee we were to pay was calculated. When we had paid our money we were given a time and date for our transit. We found the officials to be totally reasonable and helpful in fitting our desired date to their possible dates. The start time was dictated by them. It is usually pitched between 0400 and 0800 and seems to be dependent largely upon the speed your vessel is expected to maintain. They like to get yachts clear of the canal before dark.

If all goes well the transit will take from 10 to 14 hours. There are three lock chambers that lift you in turn until you reach the level of Gatun Lake. The first chamber opens into the second but the third chamber is a short run away. At the other end you are lowered by three similar lock chambers. Between the two highest locks there is a 28 mile stretch of water which is made up of Gatun Lake, a flooded valley, and a number of cuttings with romantic names like Culebra Cut, Pedro Miguel Cut and Galliard Cut, this last cutting actually splits a mountain in two.

You will be required to have a crew of five aboard, four line handlers and the skipper. In addition you will have a Company man with you. His title is 'Adviser', presumably to distinguish him from the Pilots who take the big ships through. Don't let the title fool you, he will be competent and you ignore his advice at your peril. If you follow his advice and there is an accident you will not be held to blame. If you deliberately deviate from your Adviser's instructions and trouble follows, it's all down to you.

Naturally, knowing your vessel intimately you could conceivably find yourself in the position of having to reject the advice given but I think that would be an unlikely event. I have never heard a yachtie complain about the seaman-like qualities of their Advisers. Certainly the Advisers who supervised the three transits in which I have been involved were very competent men.

Once you know when you are to take your boat through, you must start recruiting the additional hands you will need. Most yachties offer their services to other vessels as line handlers so that they can experience at least one transit as crew before taking their own boat through. There are locals who will serve as line handlers for a fee and in 1985 they were asking for thirty dollars a day. Occasionally U.S. Army or Navy personnel offer their services; be wary especially if they are not experienced yachtsmen. A Navy man might be an excellent stoker or storekeeper but that will not make him a seaman and you will need seamen for this job.

You will of course be required to feed and water the crew for the trip. At the end of the day all hands will feel as if they have done a hard day's work. Despite the work and sometimes the stress involved, most yachts contrive to make it a picnic and an occasion as indeed it should be; after all, how many yachtsmen can claim to have taken their own yacht through the Panama Canal?

Four 100-foot lines will be needed and a large bowline must be worked in one end of each rope. 100 feet is just comfortably long enough so don't try to use shorter lines. They must be strong and in good condition as they will be subjected to very heavy loads at times.

When you have dropped your lines at the Panama Canal Yacht Club it will take about twenty minutes motoring to reach the first lock chamber. Here, one of three situations will be in store for you. Most commonly you will follow a large ship in and take up your station about fifty yards astern of it.

As soon as you are positioned, four heaving lines will thud down on to your deck. Each line must be attached to one of your mooring lines, two forward and two aft. Once your pilot signals that it is in order to haul away, all four mooring lines will be taken up to the lock side and the bowlines dropped over their bollards. Any slack should be taken up by the line handlers who must stay alert because from the moment the water starts to enter the lock it will be largely their efforts that will keep the boat safe. Every scrap of slack line must be taken up without delay as your yacht rises on the incoming water. Your boat's engine must be kept running in the lock chambers as it will be needed to control your vessel when the water starts to surge in.

The inlets in the floor of the lock are enormous and an immense amount of water enters the lock in a very short space of time. The turbulence this creates is considerable and can be the source of some difficulty. Success centres on maintaining your position in the lock and this is done by use of the engine and by the prompt efforts of the line handlers. Follow your Adviser's instructions and he will carry you through what is the most difficult phase of the transit. As you reach the upper level so the world calms down and peace descends.

Don't relax .... there may be more to come!

Usually the ship ahead is towed out of the chamber by the six electric locomotives that have lines out to it and this will cause the following yacht no problem at all. Just occasionally the ship ahead will use its propellers to speed up the initial movement. Since the vessel is stationary and all those thousands of horse-power are transmitted to the water by the rotating propeller you will find yourself in the grip of a veritable maelstrom.

This happened to us and it was by far and away the worst part of the time we spent in the locks. *Didycoy* seemed to be thrown in every direction at once despite the engine running full-out in an attempt to counter the fantastic wash. Why the lines did not part I shall never know. If even one line had parted I believe the boat would have been smashed into the massive lock gates with serious, maybe total damage. Injury and quite possibly loss of life would have followed. I was both frightened and angry, as indeed was our Adviser.

The very fine British cruise-liner submitted us again to this callous treatment in the next two locks despite our Adviser's repeated requests for consideration. Although we could hear their VHF conversations with the lockmaster etc, and although we were able to speak to the lockmaster through our Adviser's radio hand-set, the liner ignored our attempts to contact her.

The locks on the canal are built in pairs so that two ships can be put through in adjacent locks. It was our good fortune to be separated from the liner for the down locks by this arrangement. We at least had the satisfaction of hearing a very courteous American Master of a freighter making his views known to the lockmaster of the last down chamber over the behaviour of the same cruise ship when it seems that a collision was avoided only by the freighter dropping her anchor in Gatun Lake.

The restrained comments of our Adviser and my own experience in two other transits of the canal suggests that our treatment was, to say the least, unusual.

Just occasionally a yacht is lashed to a tug in a lock and this can be good as the tug does all the work and stabilises the yacht. If you find yourself in this position on no account let the tug take you into the next chamber. A yacht was lost some years ago when the tugmaster apparently forgot that the boat was alongside and smashed it against the lock wall as he entered the second chamber.

We have seen correspondence in an American yachting magazine on this subject, including a letter from the Panama Canal Company, so there is no doubt that it happened. It is also on record that the Company made a generous settlement. The safety record of the canal is so good that this was the only story of serious damage to a yacht that I could unearth.

Having reached the level of Gatun lake most Advisers don't object to the sails being raised if the wind serves but will probably want you to motor-sail to maintain his schedule. If he is agreeable, he may take you through one of the smaller channels like Banana Cut, which are much more interesting than sharing the main channel with the big ships. I'll not dwell on this aspect of our passage, as part of the pleasure is your own discovery of it all.

The three down-locks are nothing like so turbulent as the three up-locks. When the gates to these locks are opened there is an outflow of water that causes a little turbulence but nothing on the scale of the up-locks. When you leave the last lock and enter the Pacific you will make your way for some thirty minutes to the Balboa Yacht Club to drop your pilot and line handlers .... the only thing they don't make a charge for. The Balboa Yacht Club will provide diesel or water to members only and to join the club would make your diesel the most expensive you have ever bought! The message is, stock up with both before you leave Cristobal Colon.

Since our transit it has become more usual for several yachts to be put through rafted up in groups without the presence of a ship, an altogether more comfortable procedure. The cost in late 1990 was little changed from the 1985 figure.

Ten miles away is an anchorage at Taboga Island, which is where most boats head for. From Taboga there is a cheap daily ferry to take you to Balboa and Panama City should you wish to visit those places. Taboga is a lovely little island with no motor traffic and the most gorgeous pineapples we have ever eaten. Shopping on Toboga is very limited so you would be well advised to stock up in Cristobal Colon before you enter the canal otherwise you will have to carry it back from Balboa or Panama City. There is neither diesel nor water available for the passing yachtie at Toboga.

Now that you are in the Gulf of Panama don't fail to visit some of the Las Perlas Islands that are just thirty miles from Toboga. Many of them are unspoilt and a number of them are uninhabited. We rate some of the bays in the Las Perlas group as amongst the most beautiful in which we have had the privilege to drop our anchor.

## The Suez Canal
The Suez Canal is an almost straight sea-level canal. It is very busy but presents no real navigational difficulties for the yachts that take passage through it. From Suez to Port Said, the canal is almost 100 miles long. As this is more than the average yacht can cover in one day most boats are required to anchor overnight in Lake Timsah at Ismailia.

The Authorities require you to employ a pilot and the first one will leave you when you anchor at Ismailia with a fresh pilot joining you for the second half of the transit the next morning. These pilots are not of the same standard as those who are employed by the Panama Canal Company, but then the job is not so difficult. Certainly it pays to keep a careful eye on the way your boat is handled throughout the length of the canal.

It is possible to complete your own paper work but it is not a task to be undertaken lightly. I have met yachties who did their own documentation and it took them three months to complete it. It must be remembered that you are operating in a land where 'baksheesh' is part of the way of life and where officials can retreat into a language that is incomprehensible to you. When I say that 'baksheesh' is a part of the way of life, I mean exactly that. It is not normally a form of corruption.

Egyptian civil servants are very poorly paid and they are forbidden to take a second job to enhance their pay. To bring their income up to a reasonable level they expect to be given a small tip for completing certain tasks. This is where an Agent earns his fee. He knows who to pay and just as important, how much should be paid. To our way of thinking it spanks of bribery but it is so much part of the fabric of Egyptian society that it is really far from bribery.

While we were in Port Said our agent took me to buy some rope. The owner of the shop supervised the purchase but an employee took the line from the reel, measured and cut it and put it into our agent's car for me. When I had paid for my purchase there was an uneasy silence as I waited for our agent to show signs of leaving. Eventually he said, 'You must give a little baksheesh to the man who did the work, as that is how he makes his living'. As soon as I had parted with what I thought was a suitable sum of money, full *bon homie* was restored. In the car the agent's first words were, 'You gave him far too much'. I felt that I couldn't win.

In 1988 we paid $165 (US) for our Agent's services in respect of our documentation and the cost of using the canal. This was for a transit from South to North. For some reason which I never did understand, the charge is a little higher if the transit is from North to South. An agent is only required for the end of the canal at which you start. In fact we paid rather more than the sum mentioned above. The extra cost was for providing and supervising a first class mechanic and a number of other services which were specific to our particular situation. I have to say that I entered Egypt expecting to be robbed at every turn. The absolute reverse was the case. We were treated with courtesy and honesty wherever we went. It is true that at the site of the pyramids at Gizah, there were touts and salesmen who had an inflated idea of the value of their services and wares but it was no worse than at tourist spots anywhere else in the world. It is also true that there was just one Customs officer at Suez who demanded cigarettes and whisky. His demands were refused whereupon he started out to make life difficult for us. Our agent asked us to leave things to him and he worked back-stage and put things right in a fairly short space of time.

Our agent at Suez was a mine of useful information, extremely helpful and completely honest. In our ignorance we had chosen well because it was obvious that some other agents charged more and gave rather less in return.

In Port Said, at the northern end of the canal, we were approached by Nagib Latiff, who said that he was an agent but that we did not need an agent as we had paid for one in Suez. However, he would be at the mooring compound each day and should we have the least problem he would come to our aid without making a charge. In the event he did much more than this and became a firm friend.

The pilots will take advantage of any yachtie who will allow them to. They can be an absolute pain with their importuning for presents which will continue undiminished no matter how much is given. We had checked with Abdul Menheim before we left Suez, on just what was a reasonable sum to give the pilot in 'baksheesh'. He had said that $5 would be reasonable and $10 would be very generous and this is what we had determined to give them. About three hours before we reached Lake Timsah our pilot started to ask and then demand that we should give him some presents. I quickly subjected him to a powerful blast of what the navy chooses to call 'Power of Command' and he said no more. When he left we gave him his $10 and he went off like a dog with two tails.

Next morning the new pilot arrived and started by telling me to give the crew of the pilot launch a packet of cigarettes each. He got his statutory earful and we had quite a pleasant trip. At Port Said he was given his present and once more we had a happy man saying good-bye to us. I sympathise with the pilots; to be habitually underpaid must be pretty tough on a man who has a family to care for. Keeping in mind the 'baksheesh' tradition of the country, I feel that it would be quite wrong to send them away without some acknowledgement of this situation.

The other side of the coin is that many yachties encourage the pilots to degrade themselves in this way by the sheer extravagance of the largesse they dispense, often throughout the length of the canal and by their unwillingness, or perhaps it is inability, to stop the importuning as soon as it starts.

# Chapter 25
# Clearance

When a vessel arrives in a foreign port there are certain formalities that must be dealt with. The whole business has its origins in the way commercial shipping is treated and one sometimes wonders if the port authorities in some places have noticed that a pleasure yacht is different from a ship.

The work involved in 'clearing in' ranges from simple and painless to frustrating, expensive and time-consuming. When you wish to leave it all happens again as you 'clear out'. I am quite sure that the multiplicity of forms we completed for various Customs and Immigration Authorities and sometimes the National Guard, the Port Medical Officer for Health and the Port Captain, were filed and will never see the light of day again until they are taken out many years later to make room for ever-more forms.

Questions like, 'what is your draught aft?' followed by 'what is your draught forward?' must at least appear to be taken seriously and an unequivocal answer given with every sign of assurance. So long as it is plausible, it does not matter if your answer is wrong. What does matter is that an answer is provided for the Port Captain to write in the appropriate space on his wretched form.

In one West Indian island, I completed four copies each of two forms (no carbon paper of course) for the Customs Officer and was told to come back later as the Immigration Officer was not available. On my return, the Customs Officer I had dealt with earlier moved to the part of the counter marked 'Immigration' and without a trace of a smile handed me four more copies, each of the two forms, that I had earlier completed for him when he was posing as the Customs Officer! On these occasions you just have to maintain a dignified silence.

On one Polynesian island, I had to return to the boat to get the engine number, something we had never before been asked for, nor have been subsequently. It was two and a half miles to the Customs post which meant that clearing-in that day required me to walk ten miles, not to mention rowing back and fore between shore and boat. It is no wonder that passage makers are a healthy lot. Our engine number is now penciled-in on the ship's papers, too late as usual!

We were agreeably surprised, and somewhat humbled, to find that someone spoke English no matter where we called. English and goodwill will usually get you what you need. If you can muster a smattering of Spanish and French you will be well away. Of those two languages we found Spanish was the more useful. We always made a point of learning a few words of greeting and courtesy in the language of our port of call and its use always seemed to please the local people even though they could speak English.

Although the questions on some of the forms we had to complete were in the language of the country, answers given in English were always acceptable. Crew lists were always wanted, usually in triplicate but in some places even quintuplicate. The information required was always the same. The boat's name at the head of the paper followed by details of each member of the crew including the skipper. The details were, fore and surnames, address, passport number, the date and the place at which the passport was issued and it's date of expiry.

It is a wise precaution to have the relevant pages of your passport photocopied. Should you lose your passport, then the photocopies should help speed up the issue of a new one.

Memorise your passport number. In extremis, it might just save you a lot of pain and aggravation.

We met a young English couple whose boat broke up in the Bay of Biscay. They were picked up by a Greek ship that took them to Gibraltar, where authority treated them with great suspicion and downright hostility. They had no papers, which was not surprising as they had to swim to the freighter that picked them up. They were repeatedly asked for their passport numbers which they could not supply. For a week or ten days they were incarcerated in unpleasant circumstances and were repeatedly and aggressively interviewed. Identification was finally established because one of them remembered their National Insurance number. Meanwhile the freighter was not allowed to leave in case it was required to take them away. As the skipper of a yacht you are in exactly the same position as the captain of that freighter.

If you have a crew member who, in a foreign country, wishes to leave your boat and claims that he does not have the airfare with which to return to his country of residence, then you as the master are responsible for his repatriation. It pays to look at potential crew members you may be tempted to pick up en route with this thought in mind.

Passports and ship's papers are not usually kept by the officials you deal with but we did find that the Harbour Master in the Galapagos retained both and in Oman, Aden, Hodeida and Curacao, Immigration kept our passports for the duration of our stay. With the exception of Hodeida they were freely available to us if we needed them to facilitate the withdrawal of money from a bank. There is no point in attempting to refuse to allow them to hold your papers. It is the law of the land and if you wish to go there you must accept their rules. Asking for a receipt will not produce one and can only create hostility. In effect the retention of your passport is a substitute for a visa and it often allows you the freedom to visit a country that is normally closed to tourists.

With the exception of some South American countries, most Customs and Immigration officials were scrupulously honest but just occasionally one would try to extract a bribe. One Customs Officer in a South American port was notorious. Because of his dandyfied style of dress he was known as 'The Cowboy'. He fined one American yachtie fifty dollars for sailing single-handed and then allowed him to continue to sail single-handed. Others were required to pay for his attendance or because they had arrived on a Sunday. He demanded one hundred dollars from us to pay for his taxi fare .... he arrived in the Customs truck ....the Customs post was a five minute ride away. This, I might say, was after he had torn up our entry papers that we had obtained at another port after four days hard work. My wife, who speaks Spanish well enough to tell him he was a thief, sent him on his way with the bus fare.

In South America you do not have much defence against this sort of thing. The system is built around bribery and corruption and you are a novice at the game. If you refuse to pay you could find yourself sitting in a corridor day after day waiting for that rubber stamp. The officials are on their home ground and you have no one to appeal to. What I do think was surprising was the fact, that in the circumstances, so many officials were completely honest.

The four days it took us to get our initial clearance into that country was entirely due to the fact that various men with the authority to use a particular rubber stamp were somewhere else when we needed them. Add to this a touch of the *manana* syndrome and you will find that four days to satisfy The National Guard, The Customs, Immigration, The Port Medical Officer for Health and The Port Captain is not excessive.

To set against that performance, it took us ten minutes to clear-in at Bonaire in the Dutch West Indies. In Oman I was sat in a comfortable chair, the fan was directed at me and then I was asked if I would prefer tea or coffee before we started the official business of clearing-in. If you want it all to be just as it is at home, why bother to sail away into the sunset?

Officials were always immaculately dressed and almost without exception very courteous. I suspect that they resented in varying degrees, the appearance and behaviour of skippers who did not measure up to their own high standards. And who should complain about that?

Before we left home we had a boat stamp made up. It was of little value until we reached Asia and Arabia where we were expected to have one. We got the impression that it was perhaps because, although these people could speak English, often quite well, they were not able to read written English and a picture (the boat stamp) on each of our documents helped them collect and collate our paper work. Most officials had sufficient English to guide us through the form filling but if you do meet any who are unable to help you the following table may be of some help.

| English | Spanish | French |
|---|---|---|
| Boat's name | Nombre de Yate | Nom du yacht |
| Country | Matricula | Ensign |
| Port of registry | Puerto Asiento | Port d'Immatriculation |
| Tonnage | Tonelado Asiento Peso Neto | Tonnage |
| Reg.Number | Numero Asiento | Numero Enregistrement |
| Length | Eslora | Longueur |
| Beam | Manga | Largeur |
| Draught | Calado | Profondeur |
| Description of the boat | Description de barca | Description du Batteau |

This will refer to the rig and the material from which the boat is built.

| English | Spanish | French |
|---|---|---|
| Colour | Color | Couleur |
| Sails | Vela | Voile |
| Hull | Casco | Coque |
| Masts | Mastil | Mat |

(This is asking for the number of masts in the boat)

| English | Spanish | French |
|---|---|---|
| Engine | Motor | Moteur |
| Engine number | Numero Motor | Numero d'moteur |
| Last port of call | Puerto ultimo-precidente | Dernier port d'attache |
| Next port of call | Puerto proximo-destino | Prochain port d'attache |

# Index